Careers in

AGRIBUSINESS
and INDUSTRY

Third Edition

Careers in

AGRIBUSINESS
and INDUSTRY

by **ARCHIE A. STONE**
   MARCELLA L. STONE
   HAROLD E. GULVIN

THE INTERSTATE
Printers & Publishers, Inc.

Danville, Illinois

Library of Congress Catalog Card No. 78-71477

ISBN 0-8134-2073-3

PREFACE

Careers in Agribusiness and Industry is written for young people who are thinking about their futures—about their careers and their work in the years ahead. They are beginning to look for answers to the question, "What am I going to do when I finish college?"

Perhaps this book will be helpful to students in high school and to those in the early years of college.

This is not a book about farming. Rather, it tells what happens to farm products after they leave the farm. It tells about the industries and services and professions that transform raw farm products into finished food and fiber products. And it tells how such goods produced in America are merchandised and distributed throughout our nation and all the world.

It tells about the industries that furnish our farmers with machines, equipment, and supplies that make possible our bountiful harvests.

And it tells how colleges of agriculture and colleges of business prepare their students for positions as managers, executives, and administrators and for other responsibilities in this essential business.

Farm-reared youth who cannot continue farming (and about 80 percent of them cannot) may have some special qualifications for careers in agribusiness and industry. Nevertheless, city-bred youth already outnumber farm youth in many of our colleges of agriculture. There they are preparing for careers in agribusiness and other professions related to agriculture.

Graduates have found ready employment. Many now serve in responsible positions. More are urgently needed. Some colleges have reported several employment offers for each graduate.

But planning *your* future and choosing your career deserve your most careful thought and study. Finding a work that you like is quite important. It's hard to advance very far unless you like your work.

You will need lots of information, counsel, and advice about many different career opportunities. The more information you get, the better your chance of making a wise decision concerning this vital personal problem. Perhaps this book will help you determine whether or not agribusiness or industry would interest you.

Choosing a career carefully and wisely and preparing for it thoroughly will put you on the road to success.

For more than 12 years Archie A. Stone was privileged to travel through all our mainland states, visiting farms, factories, and colleges of agriculture and business administration, and for 2 years prior to that, he was engaged in similar work in foreign lands. The desire to tell youth of career opportunities in agribusiness and industry came because of those travels and conferences with farmers, educators, and business executives. Their suggestions, encouragement, and stimulating support made this book possible.

Prior to the death of Archie A. Stone, the publisher of this book requested a third edition. Mr. Stone's wife Marcella asked Harold E. Gulvin, a coauthor with Mr. Stone on another book, to assist with the revision. He did and you now have the third edition. We wish you well!

Special thanks are given to Diane Gulvin for some of the sketches at the beginnings of the chapters, to Professors Walter Laramie and Stanley Smith for their valued assistance, and to all others who have helped in any way.

MARCELLA L. STONE
HAROLD E. GULVIN

ABOUT THE AUTHORS

Archie A. Stone, who wrote the first and second editions of *Careers in Agribusiness and Industry*, was formerly Head of the Department of Agricultural Engineering, Long Island Agricultural and Technical Institute. A farm machinery consultant, he was coauthor of *Machines for Power Farming*, published by Wiley. Mr. Stone, who was a world traveler, died in 1977.

Marcella L. Stone, wife of Archie and co-worker with him on all of his writings, is coauthor of this third edition. Mrs. Stone is a graduate of the University of Minnesota.

Harold E. Gulvin, also coauthor of this edition, is a consultant with Agway Petroleum Corporation. He was formerly a vocational agriculture teacher, an extension and research engineer, and Head of the Department of Mechanized Agriculture, University of Rhode Island. He is the author of *Farm Engines and Tractors*, published by McGraw-Hill, and the coauthor, with Mr. Stone, of *Machines for Power Farming*.

CONTENTS

INTRODUCTION

This book tells about careers in the business and industrial enterprises that produce, process, and distribute food and fiber to consumers—everywhere. This is a vast undertaking. During the decade of the 80's, it will provide food and fiber for our nation and for much of the rest of the world. It is supported and serviced by hosts of kindred agencies—professional, educational, financial, and commercial. For convenience, and with good reason, we shall consider all these interests as participants in *agribusiness*.

Since the first edition (1965), changes and trends have occurred that make American agribusiness a truly global enterprise. There are more people to be fed—35 million more in the United States. Our world family has grown by 100 million each year, and this rate of increase doesn't decrease. World population now exceeds 4.2 billion.

People in many countries flock to the cities, complicating the problem of food transportation, distribution, and marketing.

We have fewer but larger farms. In the developing new nations, this marks the beginning of a shift from subsistence farming to commercial agriculture. And commercial agriculture cannot exist without the support of agribusiness and industry.

Exports of U.S. farm products have greatly increased through the channels of agribusiness.

American business firms have invested in food production enterprises in foreign countries, often forming joint ventures with foreign companies.

The Agency for International Development, the World Bank, and regional banks in many countries have financed vast projects to increase food production.

The research work of our great foundations and the projects of our universities in foreign lands have brought substantial improvements and progress.

Such recent developments have expanded career opportunities. There is greater demand for graduates in business administration, agriculture, agricultural engineering, and home economics.

More women are being employed by business and industry.

Curricula have improved because of closer contacts and better working relations between industry and the colleges.

Modern business techniques and methods, data processing, and decision-making require highly skilled, better educated personnel.

FARM ⇌ INDUSTRY ⇌ BUSINESS

The farm produces food and fiber. Industry produces supplies and equipment for the farm or ranch as well as processing the food and fiber for business. Business moves the packaged food and apparel to the consumer. The whole process can be called "agribusiness." About 16 to 20 million people work in agribusiness.

What Is Agribusiness? It is the production of food and fiber: processing them and getting them to the world's people. Farmers and ranchers produce food and fiber, industry processes them, and business distributes them. Many services are needed—transportation, storage, refrigeration, credit, finance, and insurance. Hundreds of manufacturers furnish our farmers with the supplies and equipment they need to produce and protect their crops. Governments inspect and grade these products to insure quality and safety. Science, research, engineering, and education continually strive to improve each stage of this life-sustaining activity. Agribusiness employs millions of people throughout the world. People everywhere depend upon it for their food, clothing, and shelter. It has grown rapidly during recent decades. Today it offers excellent career opportunities for well trained young men and young women.

Now Agribusiness Must Grow Even Faster: Much more food and fiber will be needed. Here at home we shall have to produce, process, and distribute one-third more to feed and clothe our 240 million people in 1990 than we did in 1970. And in 1990, we shall export twice as much to our foreign customers. But we can feel confident that America can meet many of these increased needs because our agricultural research is active, effective, and broadly based.

Research Is Part of Agribusiness: Here at home, agricultural industries work with our state and federal governments in research. Over half of the research funds are supplied by industry, about one-fourth by our states, and one-fifth by our federal government.

FIGURE 1. Each year we have 100 million more mouths to feed. (Photo by Archie A. Stone)

With our agriculture supported by scientific research through the cooperation of business, industry, and government we need fear no scarcity of food in America in the decades ahead.

But the world food situation is not as bright; it is desperate.

Our World Family Is Hungry: And it is growing larger, with 100 million more members each year. Today, about one-third of the world's people do not have enough to eat. They need more food now and will need immensely greater quantities in the future.

You may think it strange that the hungry countries are usually

FIGURE 2. Primitive methods of farm-to-market transportation. Upper left: Ducks swim to market. Upper right: Donkeys carry 150 pounds of rice. Lower left: Grain goes on a carrying pole. Lower right: Hogs go on wheel-barrows. (Photos by Archie A. Stone)

those with the highest percentage of their people in farming. This is because their farmers are still close to subsistence farming, producing relatively small quantities, mostly for their own families. They have not yet become alert to the profit possibilities of commercial agriculture. They lack the aid of agribusiness which in advanced nations purchases the farm produce, adds value to it by refining and processing, and then transports it to active markets. They have little contact with the industries that supply fertilizers, chemicals, and equipment that make each acre yield more. They have little to sell beyond the needs of their families. And that little may be toted on a carrying pole or a donkey's back to a nearby village market with few customers. Two-thirds of the world people live in the food-deficient countries and a still larger proportion will be there in 1985.

The Hungry Nations Must Help Themselves: They must change to commercial agriculture. Their need for more food must be met largely by increased production within their own lands.

And they need general development, population control, more education, and more self-help while being assisted by more favored nations. It would be difficult to exaggerate the enormity of the food and fiber problem of the world family—a family that may have 6 billion members by the year 2000.

Can the Earth Then Feed Its People? To do so will require great changes, rapid progress, effective research, education and training, and great contributions from all the participants in agribusiness.

Agribusiness Will Help: A major change needed in the hungry countries is the development of commercial, market-minded food producers. In our advanced countries, commercial farming and agribusiness developed side by side; one complemented the other.

Agribusiness has many functions. It allies the farmer with thousands of business units and trade associations that find wider and better markets for his products. It brings to farmers the contribution of research and the guidance and counsel of trained extension workers; it furnishes farmers with materials and equipment that bring higher yields of better varieties. It brings the services of the professions—economists, agronomists, veterinarians, horticulturists, and animal breeders.

Agribusiness is concerned with all stages of food production from seedbed preparation through planting, cultivating, and harvesting; with refining, processing, and distribution; and with wholesaling and retailing until the product reaches the consumer.

All of these activities afford career opportunities—at home and abroad.

Helping the Developing Nations Is Good Business: It is good business to help farmers in backward nations increase their income. Then they can buy more from the advanced nations.

American Agribusiness Serves the World: Many nations buy farm products from us. We aid more than 100 nations through gifts, loans, credit, and concessions. We hope to alleviate hunger among their people and also hasten the day when they, too, can become commercial customers.

Our farm product exports are big business. They increased rapidly during the 1970's. Today, farm product export business offers many career opportunities.

FIGURE 3. Agribusiness uses fast farm-to-market transportation. Upper left: By *motor truck*—inspection en route. (Who are the inspectors?) (Courtesy, International Harvester Co.)

Upper right: By *rail*—115-car train used by Cargill, Inc., to transport grain from Illinois to Baton Rouge for export. (Courtesy, Foreign Agriculture, USDA)

Lower left: By *sea*—United States tanker *Manhattan* loads 100,000 tons of wheat for India. (Courtesy, USDA)

Lower right: By *air*—cut flowers flown by National Airlines from Florida to New York. (Courtesy, USDA)

Today, agribusiness is, indeed, a global enterprise. And it will often summon our college graduates for work in foreign lands. You will find more information on exports in Chapter 16.

But Most of the Opportunities Are Here at Home: There will be positions in our great metropolitan regions, in cities and suburbs, in towns and villages, and in rural areas. Great numbers of trained business specialists are needed. Here in our own country, career opportunities are almost limitless.

In your planning of your life work, agribusiness and the industries serving it deserve your careful study.

Chapter 1

AGRIBUSINESS—
OUR BIGGEST BUSINESS

AGRIBUSINESS—
OUR BIGGEST BUSINESS

"The future belongs to those who prepare for it." If you agree with that familiar statement, you will also agree that you can't start too soon to prepare for your future career.

If you are in high school, your most important problem—and perhaps your greatest opportunity—is planning for your future. You can choose your high school subjects and plan for the work you want to take in college.

Certainly, you can't begin too soon to think about these vital personal concerns. No doubt you will want to get a great deal of information about career prospects in the many occupations and professions that seem interesting to you. The more information you have, the better your chance of making effective plans for your higher education.

Perhaps you are "business-minded" and look forward to a career in business. If so, you will be interested in learning about agricultural business, or "agribusiness" as we shall call it.

WHAT IS AGRIBUSINESS?

Agribusiness is the world's biggest and most essential business. Everyone everywhere depends on agribusiness—all who eat, all who wear clothes, and all who live in houses. It's hard to define agribusiness in a few words because it's so big and so broad and so vital to people everywhere.

We will not include farming in our descriptions of agribusiness, even though that's where it really begins. But we will describe the things that business does to the products of the farm—

3

all the activities needed to get such products ready for you and to get them to you and millions of other consumers. You will learn how $100 billion becomes $250 billion—how $100 billion of raw food and fiber products are processed and refined and distributed by agricultural industries until they sell for $250 billion.

As you read this book, you will learn:

> What agribusiness is.
> What people do in agribusiness.
> Who employs them.
> What they earn.
> The prospects for progress and advancement.
> Of chances for a business of your own.
> How to prepare for a career in agribusiness.

The story of agribusiness, its opportunities and challenges for young people, has never been fully told. Few young persons realize the importance and vast scope of agribusiness. You will learn a good deal about it from this book, and then you will begin to see it in action in your own community. Some phase of it exists everywhere.

INDUSTRIES ⟶ FARMERS–RANCHERS ⟶
   Equipment   Food
   Supplies   Fiber

      PROCESSORS ⟶ MARKETERS
        Package   Wholesale
        Preserve   Retail

In agribusiness, transportation, technical services, and sales are made at all points. The consumer of food and clothing benefits from agribusiness.

Agribusiness has many opportunities for trained personnel. It may have a place for you. Certainly it has opportunities for many of the farm-reared youth who can't stay on the farm now that there are more large farms and fewer small ones. With professional college training you can, if you wish, return to your home community to manage an agribusiness enterprise there. (That would help reduce the outflow of talented youth from our rural, small town areas—a serious national problem.) But if you seek a career in a distant place—in a city, a metropolitan center, or even in a foreign country—agribusiness, with its many large corporations, has opportunities for you.

Of course, city-bred youth are attracted to agribusiness; many have "made their mark" in this field. A major part of its activities—especially the management and administrative functions—is in cities. Because such activities deal with products after they leave the farm, a farm background is not essential. That's one reason why you will find that city-bred youth now make up more than half of the enrollment in many of our colleges of agriculture.

Agribusiness Is a Two-Way Street

It *takes* raw products from the farm, ranch, feed lot, market garden, and orchard and then refines, processes, and distributes them to consumers everywhere.

It *furnishes* the producers of food and fiber with machines, equipment, and supplies from thousands of factories throughout our nation, together with the technical and professional services needed for efficient production.

AGRIBUSINESS IS VERY IMPORTANT

In a recent issue of the magazine *Money*, the editors rated the 10 best city areas in which to work and live through the mid-80's. In those areas, in what fields would you expect to find the best jobs? All the areas are in the so-called "Sun Belt," and the best fields to work in are: agribusiness, defense, technology, oil, real estate, leisure, government, and chemicals—and the necessary professions: medicine, dentistry, banking, and teaching. Note that *agribusiness*, which consists of growing, processing, and marketing food and fiber, is *number one*.

COLLEGES AND BUSINESS COOPERATE
IN AGRIBUSINESS EDUCATION

During recent years, our colleges of agriculture and colleges

STEEL, THE FARMER, AND AGRIBUSINESS

The farmer is the key in a multi-billion-dollar enterprise called agribusiness. It employs 40 percent of all the people in the United States, provides the nation with the highest standard of living in the world, and is built on a foundation of steel.

"In 1775 we called it 'agriculture.' Today it has become 'agribusiness'—the sum total of the food-fiber phase of our economy.

"The capital assets of agribusiness exceed in value the total industrial assets of the country."

—Dr. JOHN DAVIS

Dr. Davis, former assistant Secretary of Agriculture and former executive secretary of the National Council of Farmer Cooperatives, now directs the program in agriculture and business at Harvard University.

STEEL
COAL
CHEMICALS
RAILROADS
OIL
ELECTRICITY
AUTOMOTIVE
MANUFACTURING
CONSTRUCTION
LUMBER

MEAT PACKERS
DAIRY PLANTS
CANNERIES
FLOUR MILLS
PRODUCT REFINERIES
TEXTILE MILLS
PAPER MILLS
TIMBER MILLS

MEDICAL
LEATHER
MEAT
FRUITS
BAKED GOODS
DAIRY
FURNITURE
CLOTHING

FROM INDUSTRY . . .

American industries, like those above, supply the farmer with machines, buildings, tools, feed, fertilizer and other supplies—$70 billion worth of them.

TO THE FARMER . . .

The farmer puts his knowledge, capital ($600 billion worth), and labor to work, making wide use of steel machines to produce food and fiber valued at $100 billion.

TO THE PROCESSOR . . .

The fruits of the farmer's labor go on to the primary processors to be stored, processed, and merchandised. The worth of this service amounts to $150 billion.

TO THE CONSUMER . . .

Out of plants and mills into the marketplaces flow the finished products in a wide variety of useful forms for which buyers pay more than $250 billion.

FIGURE 1-1. Agribusiness—from industry, to the farmer, to the processor, to the consumer. (Courtesy, American Iron and Steel Institute)

of business administration recognized the growing importance of agribusiness and its need for specially trained young people.

At the same time, business and industry sensed an increasing need for college-trained employees who might become managers and executives. Now, colleges and business work together in developing specific curricula. Advisory committees from business take an active part in organizing courses. They establish close working relationships with the educational institutions. And the colleges help businesses organize "in-service" training courses for their employees. These "in-service" programs give new employees knowledge of particular methods and processes used by employers and thus prepare new employees for advancement.

College students—both men and women—are taking advantage of this new opportunity for specialized business education. One midwest college of agriculture offered its first agribusiness program in 1959. That first class numbered only 6; today over 100 are enrolled.

Yet, even today the importance of agribusiness and the concept of it as the most essential function in our whole economy are not fully realized. You will be amazed to learn how big agribusiness is, how many it employs, and of the great variety of promising careers that it offers.

HOW BIG IS AGRIBUSINESS?

The following gives the approximate number of workers in some agribusinesses:

Agribusiness	Number of Workers
Meat and poultry	315,000
Dairy	187,000
Baking	254,000
Canning, freezing, curing	227,000
Cotton mills	152,000
Seeds, feeds, supplies	2,000,000

Agribusiness employs one in five workers in private industry, or 40 percent of both private and public employees.

AN OVERALL VIEW OF AGRIBUSINESS

You can consider the food and fiber industries as its foundation stones, but agribusiness includes much more. Probably it will never be better described than it is in *The Concept of Agribusiness:*[1]

> [It includes] several thousand business units—each an independent entity, free to make its own decisions. In addition, [there are] hundreds of trade organizations, commodity organizations, quasi-research bodies, committees, and conference bodies—each largely concentrating on its own interests, which include education, promotion, advertising, coordination and lobbying.
>
> The United States is a part of agribusiness as it engages in research, regulation of food and fiber operations, or the ownership and trading of farm commodities.
>
> The Land-Grant colleges with their teaching, experiment stations, and extension functions are an integral part of agribusiness.
>
> In brief, today agribusiness exists in a vast composite of decentralization entities, functions, and operations relating to food and fiber.

Agribusiness Is Important in Every One of Our States

Take Minnesota, for example:

> The job of supplying food and fiber to a growing state and nation involves a large number of Minnesota businesses. Aiding the farmer in this task is a team of specialists ranging from manufacturers of farm machinery to the retail food distributors. This great agricultural complex, including firms manufacturing and supplying inputs for agricultural products, farmers, firms processing and distributing farm products, and others who aid in the production-marketing process, is called agribusiness.
>
> Agribusiness is an important part of our state and national economy. On a national scale, 40 per cent of all consumer expenditures in a recent year were for products having their origin in agriculture. In that year, nearly 40 per cent of the state's total labor force was per-

---

[1]John H. Davis and Roy A. Solberg, *The Concept of Agribusiness*, Alpine Press, Inc., Boston, Mass., 1957.

forming agribusiness tasks, accounting for about one-fourth of all the personal income received in Minnesota that year.[2]

Maine sends us this message on agribusiness:[3]

Today, tomorrow and always, . . . there are big jobs ahead for young people in the *food and fiber industries.* The food and fiber field provides more jobs and a wider choice of careers than any other industry, and more of these jobs are in the city than on the farm.

Success in any work demands good planning and a careful aim at the target. The food and fiber industries offer a broad target plus rich rewards for the man or woman who is willing to attain the special skills offered by an advanced agricultural education. Find out now about your career in the food and fiber industries. The information is yours for the asking.

AGRIBUSINESS HAS A PLACE FOR YOU

Agribusiness is a big business, an important business, and a growing business. It needs more young people to take over responsibilities. It has a place for you, if you are interested and will prepare yourself for a responsible position within it. Our colleges of agriculture report that 34,000 new jobs related to agriculture become available each year; most of them are in some phase of agribusiness. That's about twice as many jobs as there are agricultural graduates. So there is plenty of room for you.

The many services and functions performed by agribusiness, although varying greatly, are interrelated. So the basic curricula now available at our colleges prepare young men and women for many employment opportunities that lead to management and executive positions.

What Kind of Positions?

You can read about these in detail in almost every one of the

---

[2]College of Agriculture, University of Minnesota.
[3]Dean of Agriculture, University of Maine.

following chapters. But here are some of the general areas of business, industry, and services that offer promising careers:

> Farm equipment
> Agricultural chemicals
> Farm supplies
> Food processing
> Sales
> Wholesale and retail distribution
> Dairy industry
> Feed industry
> Meat packing and distribution
> Government services
> Financial services
> The cotton industry
> The grain industry
> Transportation
> Export and foreign trade
> Communications, publicity, advertising
> Your own business
> Farm cooperatives
> Rural electrification

Best Jobs by Growth Through 1985 in Agribusiness[4]

BS or BA Required:

Market research workers
Computer programmers
Bank officers and managers
Personnel and labor relations experts
Purchasing agents
Statisticians
Landscape architects
Engineers: agricultural, biomedical, industrial

Special Schools; One to Three Years Post High School:

Computer service technicians
Agricultural technical writers
Floral designers
Pest controllers
Surveyors
Forestry technicians

---

[4]Source: U.S. Bureau of Labor Statistics.

FIGURE 1-2. "Goods from the earth" are the merchandise of agribusiness. They flow to our food processing and meat packing industries and finally to our retail stores. Total sales at the "farm gate" are about $100 billion per year for food and fiber. (Courtesy, USDA)

REVIEW

Reading the preceding list you see that agribusiness includes many major enterprises and services. And it is served by the sciences and the professions, by education and communications, by research and finance. It employs more people than any other business; it provides more jobs in the cities than on the farm; it requires the services of two-fifths of all our working men and women.

Total sales of farm products at the "farm gate" amount to nearly $100 billion per year. But there, such products are still raw materials, not yet ready for your use. Agribusiness refines, processes, packages, and prepares them and gets them to you when, where, and however you want them. That takes lots of doing. It raises the value of our *food* and *fiber* products from $100 billion to $250 billion. And we can add still more to our dollar measure of agribusiness. Add for the farmers' purchase of production equipment and supplies; add for the cost of many related services.

Indeed it is a multi-billion-dollar business—big, broad, and important. Perhaps you can find your future in it.

Chapter 2

A GENERAL VIEW OF
OPPORTUNITIES IN
AGRIBUSINESS

A GENERAL VIEW OF
OPPORTUNITIES IN
AGRIBUSINESS

A QUICK LOOK AT CAREERS
IN AGRIBUSINESS

What are the prospects for employment and advancement?
What kind of work would you do?
Whom would you work for?
What salary might you expect?
Whom do employers want?

We'll try to answer these and similar questions in this chapter.

Because it is so big, it takes many parts to make up the whole of agribusiness. But one thing is certain. Our colleges now have curricula and training programs that will prepare you for "breaking in" at whatever place seems most promising.

First, let's see what commercial companies do in agribusiness and what types of careers they offer.

Enterprises and Opportunities

Last year, over 300 private-enterprise firms sought graduates from the colleges of agriculture and business at a midwestern university, and representatives of similar companies interviewed college graduates in every one of our states. They wanted college-educated youth for work in enterprises such as the following:

1. *Manufacturing and Processing:* These include refining raw materials from the farm into finished food and fiber prod-

ucts and manufacturing production materials, supplies, and equipment for farmers. Many thousands of companies throughout our nation engage in such agribusiness activities.

2. *Distribution:* All the finished products must be distributed, must be brought to the consumers, and must be on hand ready for them when they want them. So we have many enterprises and activities concerned with distribution.

Distribution agencies take products from manufacturers or processors and make them readily available to us.

Transportation services include railroads, motor trucks, barges, steamships, airplanes, and pipelines. This immense transportation activity requires many special services—credit, insurance, finance, accounting—all affording career opportunities.

Wholesale or terminal markets buy, assemble, grade, package, sell, and ship farm products. They also require special business services and trained personnel.

Warehousing and storage enable suppliers to concentrate stocks at critical locations and to keep them ready for prompt local shipments.

Retail outlets include retail dealers, independent stores, chain stores, supermarkets, farm cooperatives, discount stores, and others. They buy, store, display, advertise, package, and sell. They are the final link in the chain of distribution.

3. *Services for Agribusiness:* Many specialized business services are required—commercial, professional, and scientific. Here are some of them:

Banking and finance
Insurance
Engineering
Farm management
Soil and water conservation
Storage of farm commodities
Publicity, advertising, communications
Trade association services
Farm organization services
Grain and commodity exchange services

FIGURE 2-1. Headquarters of the U.S. Department of Agriculture in Washington, D.C.

Rural electrification
Land appraisal services

4. *Government Services:* Almost all divisions of our government—federal, state, county, city, township—require specialists for work related to the "business" of agriculture.

In fact, there are so many opportunities in government service that we shall refer to these again in Chapter 14.

The U.S. Department of Agriculture is the largest employer. It provides many essential services—professional, technical, financial, educational, production control, price stabilization, quality control, and inspection. These require trained representatives in almost every one of our 3,000 counties, in cities and villages throughout our land, and in many foreign countries as well.

Each of our 50 states has its own department of agriculture, which carries on a multitude of essential services—some of state-wide importance as well as local projects or enterprises.

Every county has one or more employees who serve in the non-farm activities of agriculture.

Most government services are under the supervision of federal, state, or city civil service commissions.

5. *Professional and Scientific Services:* The manifold activities of agribusiness depend on contributions from those engaged in many sciences and professions:

Economists
Research scientists
Chemists
Agronomists
Biologists
Engineers
Home economists
Nutritionists

WHAT AGRIBUSINESS GRADUATES DO, AND WHOM THEY WORK FOR

At The University of Wisconsin placement office, students are assisted in getting positions in agribusiness. In recent years the number of jobs available has been far greater than the number of graduates. For example, in a recent year:

- 448 agribusiness organizations listed available positions with the placement office.
- 132 of these sent a company representative to meet with the students.
- Initial salaries for agribusiness graduates are above the average for other college graduates.

A recent survey by The University of Wisconsin of B.S. graduates shows:

10% entered farming
23% entered government service or education
30% enrolled in graduate study
20% entered agribusiness
8% other

Figures are approximate but comparable to those from other universities.

A follow-up study of 258 students who received B.S. degrees in Agricultural Business and Management was conducted by The Pennsylvania State University.

Here is the record of the employment of those graduates:

Type of Employment	Number Employed	Type of Employer	Number Employed
Accountants	7	Chemical companies	14
Administration-executives	22	Credit agencies—banks	9
Agricultural advisors	11	Dairy companies	9
Agricultural economists	35	Electric companies	7
County agricultural extension service workers	16	Farm co-ops	21
Graduate students	12	Farm machinery manufacturers	5
Farmers	18	Feed and seed companies	9
Farm managers	13	Hardware stores	2
Foreign service workers	3	Insurance companies	10
Inspectors	7	Meat packaging firms	2
Market specialists	8	Owner-operators	19
Owner-operators	15	Penn. State University	30
Product and quality control workers	8	Other universities	22
Purchasing agents	7	Food processors	8
Rural sociologists	4	Public agencies (state)	6
Salespersons	22	Railroads	3
Sales supervisors	16	Restaurants	1
Statisticians	6	Steel companies	4
Store managers or ass'ts	15	Trade associations	2
Territory supervisors	13	Miscellaneous	75
	258		258

You will recognize the names of some of these employers:

Chemical Companies – American Cyanamid, Charles Pfizer, du Pont, Olin Mathieson, Union Carbide.

Credit Agencies – Bank for Cooperatives, Farmers Home Administration, Federal Land Banks, local banks.

Dairy Companies – Bordens, Dairyman's League, marketing cooperatives.

Electric Companies – General Electric, Sylvania, Philadelphia Electric, Pennsylvania and Western Electric.

Farm Cooperative – Agway Inc., a processing and marketing cooperative.

Farm Machinery Companies – Ford, International Harvester, Massey-Ferguson.

FIGURE 2-2. This is the top-management group of a large regional purchasing-marketing cooperative. Most of these men are college graduates who started as trainees and who have earned their rank by taking several responsible positions to gain experience. (Courtesy, Agway Inc.)

Feed and Seed Companies – Atlee-Burpee, McMillen, Ralston Purina, individual mills.

Food Stores – A & P, American Stores, Penn Fruit.

Insurance Companies – Equitable, Nationwide, New York Life, Prudential, individual agencies.

Meat Packers – Swift and Company, Wilson.

Processors – Campbell Soup, H. J. Heinz.

Public Agencies – Pennsylvania Department of Welfare, state correctional institutions, state hospitals.

Steel Companies – Jones and Laughlin, U.S. Steel.

U.S. Department of Agriculture – Agricultural Stabilization and Conservation Service, Farmer Cooperative Service, Foreign Agricultural Service, Soil Conservation Service.

Opportunities are so abundant you need not fear unemployment if you are well qualified. And probably you can choose the type of work and the location that you think best. A recent survey among agricultural colleges revealed 1½ jobs for every graduate.

Jobs Available to B.S. Graduates in 12 States
of North Central United States

Fields of Occupations	Available Jobs
Banking and credit management	396
Communications (radio, TV, newspapers, magazines)	151
County extension work	240
Dairy plant and food processing	320
Farming and livestock management	1,240
Landscape design, city planning, greenhouse and nursery management	299
Livestock buying	122
Pest control and entomology	172
Professional farm management	197
Sales and management in business and industry	3,876
Soil Conservation Service, Farm and Home Management, other government jobs	265
Turf management	188
Vocational agriculture teaching	580
Wildlife and recreation management	161
Wood utilization and forestry management	96
Others	194
Total	8,497

There were 5,756 B.S. graduates in the 12 states, or about 1½ jobs per graduate.

What Salary Might You Expect?

The College Placement Council reports that starting salaries for college graduates have increased about 6 to 7 percent during each recent year. For graduates in general business, salaries now average nearly $1,000 per month.

One college of agriculture reported good jobs available for its recent graduates, with salaries ranging from $9,000 to $12,000 per year.

Holders of advanced degrees obtain higher salaries, around $200 more per month for master's degrees and up to $500 more per month for Ph.D.'s.

Many college students and many high school students preparing for college do not realize the promising opportunities in agribusiness. It can afford a large number of them a life's work of satisfaction and achievement. It has special attraction for the 8 out of 10 young folks who cannot stay on their home farms but must seek careers elsewhere.

Whom Do Employers Want?

Employers seek college graduates because:
1. They want specially trained people to fill specific positions.
2. They want young people who will "grow" with experience and become managers and executives.

Employers think not only of whom they need today but also—perhaps even more—of whom they will need tomorrow.

A new employee may be given a special, specific assignment. Success in a special assignment often leads to a more general assignment with greater responsibility. Beginners may be "specialists" of high or low status. But successful "specialists" usually become "generalists," and ultimately they become the managers and executives.

What Employers Look for in a College Graduate

Even though you may still be in high school, you can start preparing for college and your future career. You can develop the qualities and abilities employers want. Here are a few basic qualifications employers look for.

A Good Scholastic Record: You'll get attention from prospective employers if you are in the upper third of your class; however, employers don't judge you entirely on your marks. Far from it!

Interest in Their Business: Employers like to have you show interest in, and some knowledge—however slight—of, their particular enterprise. They want you to like it. This is something you can think about and read about in high school. Getting information on the major fields of business and the professions is a broadening experience.

Ability to Get Along Well with Others: This is most valuable. If you learn to like others, to be interested in them, you will get along well with them. This will be most helpful when you are required to manage, direct, and supervise others.

Good "Communications": That means the ability to express yourself clearly, directly, and forcefully, both orally and in writing. It also means that you should have something worthwhile to "communicate"—some good ideas or information. That comes from good listening and good reading. Both are part of communication.

Activity and Resourcefulness: These qualities are shown by your extra-curricular activities, how much you take part, and how much you contribute to school life. All such experiences develop initiative and self-reliance. Later, all such experiences will help you become a "self-starter"—a resource in great demand, if it is backed by experience and good judgment.

Curiosity: Employers know that new ideas usually come from one who has an inquiring mind. One who has new and fresh ideas is usually one who can control and develop them. New ideas are vital to the success of almost every business. It is estimated that one-fourth of the products and services we use today were unknown 10 years ago.

Analytical Ability: That's the ability to get information, to assemble and interpret it, and to draw sound conclusions from it.

We will mention other important qualities and abilities that will help in whatever career you may choose. But this one seems outstanding in importance—that is, *a liking for your chosen work.* If you find a career that satisfies and fascinates you, hard work will be easy and pleasant; you will become dedicated to it. Unfortunate, indeed, are those who don't like their work, who think it drudgery and who slave each day without enthusiasm. They never get very far.

So when you consider a certain career, keep this question in mind: *"Would I like the work?"* That's not only a most important question, but also it's a mighty hard one to answer. But in trying to find the answer, you'll see the value of getting lots of information about each career that may interest you. And that will help you make a wise choice.

This is how one large agribusiness, Agway Inc., selects its new employees:

> The criteria we use for selection are many. We take a look at how the applicants present themselves and try to determine if we feel they will fit Agway and our membership. We are very interested in persons with some experience on a farm or with some knowledge of farming. This is less important for some jobs, such as at our garden stores, but a knowledge of plants and horticulture is necessary for those jobs. We take a look at their education and the types and breadth of courses they have taken along with their majors. Courses in business and agriculture seem to give the background we are looking for.
>
> We attempt to measure their written and oral skills, their skills in developing inter-personal relationships, and their potential leadership skills including judgment, decisiveness, initiative, energy level, stress tolerance, etc. We endeavor to determine their interest in Agway and agribusiness.

You can begin now to prepare yourself to meet the preceding criteria so that as a senior in college you will be ready for your first placement interview.

In the following chapters you'll read about agribusiness in some detail.

Chapter 3

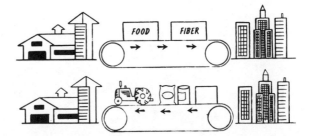

THE "MERCHANDISE" OF AGRIBUSINESS

THE "MERCHANDISE" OF
AGRIBUSINESS

A LOOK AT THE "MERCHANDISE"
OF AGRIBUSINESS

You will understand agribusiness better if we go back to its beginning. There you can see where its primary materials come from and learn how they are produced and who produces them. This will give you a valuable broad background.

The basic merchandise of agribusiness is of two kinds—first, the farm and ranch products that move through many channels to the consumer and, second, production supplies and equipment that move from factories to the farms and ranches.

Let's call the first "goods from the earth" and the second "goods from industry."

INDUSTRY → FARM → MARKET → CONSUMER

Goods from the Earth and
Goods from Industry

Goods from the earth come from our land, from our farms, and from the enterprise of our farmers and ranchers. Products from our lands are the raw materials of agribusiness. Later you will learn how these primary goods are assembled, refined, processed, transported, and distributed to every home in our nation and to many other nations of the world—a business service that almost triples their value.

But the whole process starts with the land. Today, our 50 states give us a total land area of 2,264 million acres, and much of

FIGURE 3-1. Farms adjoin our towns and cities. Looking westward from the village of Gilby, N.D. (Courtesy, USDA)

this total is used to produce the merchandise of agribusiness—crops, livestock, and forestry products. Here's the way we use our land:[1]

How We Use Our Land

	Million Acres
Cropland, including cropland used only for pasture	472
Pasture and grazing land (including woodland and forest land that is grazed)	802
Forest and woodland that is not pastured or grazed	525
Other land—outside of farms—cities, highways, parks, airports, railroads, unused wasteland, etc.	465
Total Land Area	2,264

[1] U.S. Bureau of the Census.

Look closely at these figures and you will note that crop production, livestock production, and forestry use about 80 percent of our total land. Cities still occupy a relatively small portion of our 50 states.

Some of these acres are good and rich; some are lean and poor; some are in remote wide-open spaces; some are right next to our towns and cities. But products from almost all of these millions of acres become the "merchandise" of agribusiness.

NOW LOOK AT OUR FARMS

We now have about 2.6 million of them. Each year our farms decrease in number, but acreage per farm remains about the same.

How Big Are They? They now average 400 acres each. That's the largest average size ever reported. About 140,000 of them contain 1,000 acres or more; nearly 200,000 have 500 to 1,000 acres; about 700,000 have 10 to 100 acres; and some 84,000 have less than 10 acres. Our small farms have rapidly decreased in number during recent years.

Our farms are becoming more highly specialized. More and more they depend on one principal source of income. So we can classify them according to that major source.

Livestock farms and ranches outnumber all other types. Then follow:

> Dairy farms
> Cash-grain farms
> Cotton farms
> General farms (income from three or more sources)
> Tobacco farms
> Poultry farms
> Fruit and nut farms
> Vegetable farms
> Nursery, florist, and greenhouse enterprises

What Are These Farms Worth? The average value of land and buildings per farm is $140,000. Production assets on our commercial farms average nearly $200,000 per farm. Total investment in agriculture is $671 billion. So you can see that the great structure

FIGURE 3-2. Our farmers use soil-conserving practices. Contour strip cropping on the Obert Olstad farm (front) and the Carl Backum farm (rear) near Wesby, Wisc. (Courtesy, USDA)

of agribusiness rests on solid foundation stones—our land and our farms.

Good land, good farms, and good farmers—from these come the "goods" for agribusiness.

A SALUTE TO OUR FARMERS

Each year our farmers perform a production miracle. Efficient and progressive, they give us abundant supplies of food and fiber. Since 1967, farm output per man-hour has nearly doubled.

Our farmers also control a large share of our nation's forests. Three out of four forest owners are farmers. They help produce all our forest products, including the pulpwood for the 400 pounds of paper that you and each one of us use every year.

Farmers are the suppliers of agribusiness and also its customers. They are good in both capacities, and their financial position is sound:

Their assets in a recent year .... $671 billion
Their total liabilities .... 102 billion
Their equity .... 569 billion

Most farm products are refined and processed for human food. We Americans spend about $180 billion for *food* each year, nearly $1 out of 5 of our disposable income. Yet the cost of food, based on the purchasing power of an hour of factory labor, seems to be almost as low as it has ever been.

Farmers not only *sell* to agribusiness firms but also *buy* from them. Their purchases keep factories humming.

How Much Do Farmers Buy from Agribusiness?

Each year farmers buy some $70 billion of goods and services to produce their crops and livestock. Here are some of the items on their big shopping list during a recent year:

Feed and seed .... $17.0 billion
Tractors, motor vehicles, machinery ...... 8.8 billion
Fuel and petroleum products .... 10.1 billion
Fertilizer and lime .... 5.6 billion
Electricity—33 billion kilowatts
    More than 30 percent of all used in the United States.
Rubber—360 million pounds
    About 5 percent of all used in the United States.

Farmers buy lots of automotive equipment from industry. They use over 3 million motor trucks, more than 4 million automobiles, and over 5 million tractors. Within a recent 10-year period their investment in labor-saving equipment increased from $20 billion to $40 billion.

Farmers use more horsepower (electrical and mechanical) than all other American industries. They buy 20 million tons of nitrogen fertilizer and 20 million tons of lime from agricultural chemical suppliers each year. Their purchases of irrigation equipment have increased steadily; now farmers irrigate over 40 million acres.

All these goods, and many others like them, comprise the agribusiness merchandise that flows from factory to farm.

Manufacturing production goods and supplies and delivering

FIGURE 3-3. "Goods from industry" produced in our factories flow out to our farms. (Courtesy, International Harvester Co.)

them to our farmers keep factories busy. Our farmers' income from the sale of their products, plus some non-farm income, is now about $100 billion annually. And farmers spend a total of some $70 billion. Their purchases, coupled with those of other related businesses, are so great that nearly 40 percent of our people depend on agriculture.

Agribusiness has two vital functions in our economy:

1. It receives "goods from the earth" from farmers and refines, processes, and distributes these goods.
2. It furnishes "goods from industry" to farmers so they can produce efficiently and abundantly.

For this dual function, agribusiness uses billions of dollars of capital and employs millions of workers, managers, administrators, and executives. Agribusiness includes our marvelous marketing system, which each year distributes *food and fiber* products valued at $250 billion.

Opportunities for successful careers in agribusiness are so many and so promising that your time will be well spent in making a thorough study of them.

So next let's consider how to prepare for a career in this attractive field.

Chapter 4

HOW TO PREPARE FOR
A CAREER IN AGRIBUSINESS

HOW TO PREPARE FOR
A CAREER IN AGRIBUSINESS

This chapter will tell you how to prepare for a career in agribusiness—how to lay the foundation for progress and advancement. We will discuss some fundamentals and answer questions such as these:

> What foundation stones are necessary?
> Where can you study agribusiness?
> What will you study?
> Are you eligible for college entrance?
> What are the major courses for specific areas of agribusiness?
> How can you make contacts with industry?
> What is the value of graduate study?
> Where can you obtain in-service training?

YOUR FOUNDATION WILL BE BUSINESS
PRINCIPLES AND AGRICULTURAL SCIENCE

First, you should know that the best preparation will require four years, or more, of serious college study. Although agribusiness includes many diverse activities and enterprises, our colleges have developed specialized curricula that include essential business and agricultural subjects. Also, your work in college will give you contacts with business firms, opportunities to visit their plants and factories, and often opportunities for practical experience during the summer periods. In brief, these new agribusiness curricula are designed to help you learn the principles of the science of agriculture and the fundamentals of business and economics. Such a background of formal education, when coupled later with experience, will help you toward the professional status of a business executive. You will need this broad background because you are not heading for a job as a clerk or a routine employee but for a career as a professional in business administration and management.

Today's relationship between business and agriculture was well stated in an address by T. V. Houser, formerly Board Chairman of Sears, Roebuck and Company:

> The agricultural world and the industrial world are not two separate economies having only a buyer-seller relationship. Rather, they are so intertwined and inseparably bound together that one must think of them jointly if there is to be any sound thinking about either one.

WHERE CAN YOU STUDY AGRIBUSINESS?

Among the first to recognize the importance of agribusiness were our land-grant colleges and universities. Each state has at least one land-grant institution; some of our southern states have two. There are 68 in all. These institutions are the homes of our state colleges of agriculture. The Land-Grant Act providing for their establishment through grants of federally owned land and for their financial support was signed by President Abraham Lincoln in 1862.

These new colleges emphasized practical forms of education. Before they came, higher education had been classical and restrictive. Afterwards "knowledge for use" got more attention and emphasis.

> Today the old exclusiveness is gone. Higher education has become more widely practical and more widely available. No longer the privilege of the few, it is opportunity open to all who can benefit from it and who have the will to meet its demands.[1]

At most land-grant institutions, agribusiness curricula are administered by the colleges of agriculture; at those in some states,

---

[1]The National Association of State Universities and Land-Grant Colleges, 1 du Pont Circle, Suite 710, Washington, D.C. 20036.

FIGURE 4-1. When the weather is good you can study on the campus lawn. (Courtesy, Cornell University)

by the colleges of business administration; and at several, jointly by both colleges.

Most of the land-grant institutions have graduate programs in agribusiness and economics. They conduct research and have active cooperation from business firms in their areas.

In addition to the land-grant colleges there are many other highly creditable degree-granting institutions where you may specialize in agribusiness. Good examples are California Polytechnic State University at San Luis Obispo and Texas Technological College at Lubbock. Also, the University of Idaho offers three major curricula in agribusiness: agricultural economics–agribusiness, animal science–agribusiness, and soil science–agribusiness.

Many privately supported colleges have programs that prepare students for careers in agribusiness, with especially valuable courses in agricultural economics and business administration.

Junior and community colleges in almost every state have offerings that can start you on your way—courses giving credits that will be accepted at the four-year terminal colleges. Many states have two-year courses at their technical and business institutes that give transferable credits that will be accepted at four-year colleges.

There are plenty of places to study. Your guidance counselor or high school principal will have much good advice and information for you. And you might write to the Department of Education at your own state capital.

CAN YOU QUALIFY FOR ADMISSION TO COLLEGE?

Guidance counselors can be most helpful in this respect. Their personal service to you will be of great value all through your high school years. They will aid in scheduling a program that will qualify you for college entrance and in selecting subjects that are related to your chosen career. They will help you make a careful check of the entrance requirements at the college of your choice. Here are some examples of entrance requirements.

Can You Meet These Requirements When You Graduate from High School?

Texas A & M University: Graduation from an accredited high school with 15 credits acceptable to the college is essential. If in the lowest quarter of the class, an applicant may be required to take an entrance examination. Acceptable credits include: 4 in English; 2 in social science; 2 in algebra; 1 in plane geometry; 1 in natural science; 5 electives, such as foreign language, mathematics, natural science, social science, speech.

Cornell University: Requirements include: four years of English; a foreign language; 2 units of mathematics; and graduation from an accredited high school. Applicant also must have taken the Scholastic Aptitude Test of the College Entrance Examination Board. The college reviews the credentials of each applicant and considers the quality of work in the subjects taken in addition to the background, experience, character, and personality of the applicant.

The University of Wisconsin: The applicant must graduate from an accredited high school; be in the upper 50 percent of the graduating class; and present required test score reports. Also, 16 units of study, each unit representing a full year's study or its equivalent in a given subject area, are required.

Subject	Units Required
English	3
Algebra	1
Geometry	1
Any two of these:	
Foreign language—2 units History and social studies—2 units } .. Natural science—2 units	4
Other electives	7
Total units required	16

University of California at Davis: Graduation from an accredited high school and presentation of a satisfactory score on the College Entrance Examination Board Scholastic Aptitude Test are necessary. *Subject requirements* include history, unit; English, 3 units; mathematics, 2 units; laboratory science, 1 unit; foreign language, 2 units; advanced course from one of the following: mathematics (in intermediate or advanced algebra, trigonometry, solid geometry, or other course for which trigonometry is a prerequisite), 1 unit; foreign language, 1 additional unit; science, 1 unit of either chemistry or physics in addition to laboratory science mentioned previously. The applicant must then have sufficient elective units to complete a minimum of 15 standard entrance credits.

As for scholarship requirements, an average grade of B (3.0 based on a marking system of four passing grades) is required in the first five subjects listed in the preceding paragraph that are taken in the tenth, eleventh, and twelfth years. Courses taken for credit in the ninth year need show passing grades only.

When you study entrance requirements, you will realize how much counselors can help in charting your way toward college entrance. You will need their help. Don't take admission to college for granted. It's going to become more difficult. Enrollments now are greater than expected. More high school graduates are going on to college, and still more will go on in the years ahead. Compe-

FIGURE 4-2. As a student you will sit in a class such as this one. (Courtesy, Cornell University)

tition to "get in" will increase, especially in the publicly supported colleges, which even now have about 60 percent of the total enrollment, and their proportions will increase rapidly in coming years.

CURRICULA FOR AGRIBUSINESS

Curricula for this business profession are given various titles at the different colleges. Core-type curricula are designated as agricultural business administration, agricultural business and industries, agricultural business management and administration, agricultural industries, or agribusiness.

What Subjects Will You Study?

Now let's note some of the courses included in the basic curricula at certain institutions—not a complete list, just some examples. These will give you some idea of the wide range of information and knowledge you can obtain from the formal study of agribusiness.

California Polytechnic State University: Marketing programs in California, agricultural business organizations, business credit and finance, sales and service, agricultural business management and government policy, business communications.

Purdue University: General chemistry, economics, accounting, English composition, agricultural business policies, corporation finance, money and banking, agricultural cooperation, marketing farm products, agricultural statistics.

Oregon State University: Food and agriculture, agricultural marketing, land economics, agricultural cooperation, agricultural prices, finance, market analysis, money and banking, consumers and the market, applied agricultural economics, accounting, statistics, real estate law, investments, salesmanship, human relations in business.

Kansas State University: Agricultural marketing, accounting, prices and market structure, money and banking, farm management, business law, flour and feed milling, economics, agricultural policy.

University of Tennessee: English, chemistry, accounting, mathematics, journalism, agricultural economics, business law, finance, statistics, animal breeding, agronomy, horticulture, labor economics, speech.

You Can Specialize If You Decide To

Through choice of your major and your electives or by choosing a separately organized curriculum based on the more general ones named in the preceding section, you can specialize in certain selected areas of agribusiness. In addition to certain basic and fundamental courses required of all students, you can prepare to enter definite enterprises or services such as these:

Dairy industry or dairy manufacturing
Agricultural economics

Marketing of farm products
Food processing
Banking and finance, credit
Insurance
Feed industry
Grain and seed industry
Cotton industry
Farm chemicals
Farm cooperative management
Agricultural services
Oil-seed processing
Storage and warehousing
Farm supply stores
Farm equipment industry
Land appraisal
Communications, publicity, journalism
Livestock and meat industry

All of the enterprises just listed and many others comprise agribusiness. A successful career in any one of them requires a basic knowledge of business and agriculture.

Faculty committees constantly improve the offerings to keep their programs up to date and to give the graduates opportunities in newly developed fields. They have worked out special programs that prepare individuals for employment in specific agribusiness industries and services without eliminating the required basic studies.

Following are some examples of courses and subject matter included in such special offerings.

Special Curricula for Definite Professions

Agricultural Journalism: Subjects include agricultural news writing, practice in editing, home economics news writing, advertising, publicity, media and methods, advanced agricultural writing, editing bulletins.

Food Science: This curriculum is to prepare students for leadership in such jobs in the food industries as plant operation and management, quality control, sales, and inspection. Courses include: food chemistry, microbiology, bacteriology, biochemistry, chemistry, physics, mathematics, and electives.

Agricultural Economics: This is a major professional field with many of its phases pertaining to agribusiness. Curricula include: agricultural industry, business law, field practice (on production, processing, marketing, handling of farm products), economic analysis, agricultural finance, agricultural business management, agricultural marketing, cooperation in agriculture, land economics, farm management.

Dairy and Food Industries: Curricula in these fields prepare students for careers in business, management, processing, quality control, equipment development, and sales or research and development work in many of the nation's large business enterprises. Curricula include: survey of dairy and food industries, creamery operations and management, principles and practices of cheese making, market milk, regulatory and quality standards, dairy and food mechanics, ice cream, concentrated milk products, dairy chemistry.

Feed Technology: Students are prepared for responsible positions in the feed industry, one of the largest dollar volume farm

FIGURE 4-3. The academic advisor assists in preparing a schedule of classes and helps in adjustment to college life. (Courtesy, Iowa State University)

supply industries. This is a relatively new and rapidly growing agribusiness, offering opportunities in the field of animal nutrition, administration, or operation. Curricula include: chemistry, biology, elements of milling, milling industry seminar, general physics, elements of feed manufacture, principles of feeding and animal nutrition, labor management, feed forming and blending, anatomy, physiology.

Dairy Manufacturing: This curriculum is for students preparing for positions in the processing, manufacture, sale, and distribution of dairy products. It requires a knowledge of both dairying and business practices. The dairy industry is one of the largest in agribusiness. We will study it in detail in Chapter 6. Courses in dairy manufacturing include: milk procurement, dairy chemistry, dairy testing, organic chemistry, biochemistry, economics, bacteriology, physics, judging dairy products, dairy plant operation, business law, mechanical drawing, food microbiology, dairy refrigeration and machinery.

HOW ABOUT GRADUATE STUDY?

Plan for it. Look forward to it, even though you may think four years of college is all you can afford. You can't tell what the future may hold for you. Some years hence your employer may gladly subsidize your advanced study.

Business now places a great value on higher education and gives preference to holders of advanced degrees when choosing employees. There is a growing need for professionally trained people in business management.

The following are the estimated starting salaries of the graduates of 14 north central colleges of agriculture in a recent year:

Degree Awarded	Monthly Salary	Yearly Salary
Bachelor of Science	$ 916	$10,992
Master of Science	1,117	13,404
Doctor of Philosophy	1,431	17,172

It certainly would pay you well to go for a higher degree; look at the salary differences!

Even while in high school you can begin to think about graduate study. Don't be discouraged if that goal seems far distant. It is well worth striving for. An advanced degree will bring you a higher salary, more promising opportunity, and brighter prospects for advancement.

Graduate work in one or more of the several areas relating to agribusiness is available in many of our states. Specific fields include: agricultural economics, agricultural business administration, dairy and food industries, animal nutrition, food and nutrition, dairy science, dairy manufacturing, milling industry. A properly qualified student who wants to prepare for a special field may propose a graduate study program for the approval of the faculty members directly concerned.

In general, admission to the graduate school is open to holders of bachelor's degrees from accredited institutions. But your scholarship records will be examined closely for evidence of your ability to carry graduate work.

You will have to devote more time to your formal education to earn an advanced degree—a minimum of one year of study beyond the bachelor's degree to earn a master's degree, at least three years of study beyond the bachelor's degree to earn the Doctor of Philosophy degree. This will take more hard work on your part, and your formal education will cost more. But there are many opportunities for financial assistance in the various graduate schools—fellowships, scholarships, and part-time employment opportunities.

When you are planning your undergraduate program, look ahead still further. Through all your high school days and undergraduate days, measure carefully the value and advantages of graduate study.

Then you will be ready if you can open that door to enhanced opportunity.

AFTER GRADUATION COMES IN-SERVICE TRAINING

You May Start as a "Specialist"

You may be hired for a definite and specific job. Perhaps your employer finds that your education matches a certain work assignment. If so, we might say that you start as a "specialist" even though you may have a rather limited and temporary responsibility. However, if you succeed there you may be in line for greater opportunities. You may reveal, and your employer may discover, qualities and abilities theretofore unknown to you both. So you progress from the status of a limited "specialist" toward that of a "generalist."

You May Start as a "Generalist"

Your employer is making an investment in you. Because college graduates today start at good salaries, a company feels that its investment should pay dividends. You may begin your new job following a long-range plan established by the company which will give you a general background of various aspects of the company's procedures and is intended to make you a most valuable employee of the company in 5 to 10 years.

You Will Need In-Service Training

In-service training is a most important step in your career. Your employer, whether the company is large or small, will lay out a program that will make you more important and valuable to the company. All in-service programs have that objective.

Some in-service programs start with orientation to acquaint the new employee with the various activities and departments, with company organization, with location of branch houses, with methods of distribution, and with other fundamentals of the business structure.

For your first temporary assignment, your employer may try to place you according to your personal desires and interest. Perhaps you have developed these during your college years, through summer employment, and through other contacts. But your first job assignment should not predetermine your future with the company.

A good training program will give you some understanding of the functions and activities of departments such as these:

Sales and Distribution
Purchasing
Accounting
Personnel
Credit and Collections
Price and Contract
Public Relations
Industrial Relations
Traffic and Transportation
Product Planning

Large corporations have adequate staffs in all departments to carry on the essential functions just listed. But even in small concerns, almost all of these same functions are required. Often they must all be performed by a very few persons. So you can see that managers of small businesses must be versatile and competent.

FIGURE 4-4. In-service training features first-hand contact with on-farm operations for a wide variety of new employees, ranging from persons in sales, advertising, or public relations to those in engineering and service. (Courtesy, Sperry New Holland)

Whether you work for a large corporation or a small company, you can learn much from its "front-line" managers and executives. In fact, association with seasoned, experienced administrators may be the greatest benefit of in-service training. Their counsel, advice, and example will help through the years when you may think your progress is too slow.

Agway Inc., a large regional cooperative, describes its training program as follows:

> Basically, we have two kinds of *management* training programs: one prepares a trainee to be a manager of a retail store, and the other prepares him or her to be a manager of a petroleum plant.
>
> We have no formal program for preparing an employee to be a manager of a feed mill or a fertilizer plant; however, from time to time we do conduct a more or less informal on-the-job training program in the preparation for production management positions.
>
> Our *accounting* training consists of a formal, 18-month program which prepares the trainee for a responsible position in the office of the controller or in the data processing office.
>
> We have no formal *sales* training program. Usually our college people will begin in one of our *management* training programs and then switch to "sales."

NEXT—A CLOSE-UP OF SOME AGRIBUSINESS INDUSTRIES

Now we will look more closely at certain agribusiness enterprises. But we can't describe them all because there are so many. But these few will tell you of the importance of such industries, the careers they offer, and more about in-service training.

These are major agribusiness industries:

Food industry
Dairy industry
Grain industry
Feed industry
Meat and livestock industry
Cotton industry
Farm and farmstead equipment industry
Agricultural chemicals and farm supplies industry
Ornamental horticulture

Farmer cooperatives
Rural electrification industry

These industries are discussed in detail in the following chapters.

For descriptions of positions that may be of interest to you, obtain a copy of the *Occupational Outlook Handbook* from your counselor or library. They can obtain it from the Superintendent of Documents, Washington, D.C. 20402, for about $8.00. It is revised every two years and describes hundreds of jobs.

Chapter 5

THE FOOD INDUSTRY

THE FOOD INDUSTRY

You will be better able to evaluate the many types of career opportunities in the food industry when you appreciate how extensive and vital that industry is and the great variety of operations and services it performs.

WE ALL DEPEND ON THE FOOD INDUSTRY

Ever wonder how much food you eat in a year? *About three-fourths of a ton, if you are "average."* If you're a heavy eater, perhaps you eat a ton. That's a lot of food. And it takes a lot of work and a lot of people to get it to you just when, how, and where you want it. But that's what the food industry does.

Farmers, processors, packagers, distributors, carriers, wholesalers, retailers—all help get it to you. They all are part of the food industry.

We can't put the whole story of this great industry into a brief chapter. It's too big for that. But we will tell you about some of the industry's special enterprises and activities and, especially, about the prospects for careers in it. This industry needs "youth-power." Once you get interested, you will think about the food industry and read about it; then you'll begin to see some of its activities right near you—wherever you live. It offers many careers for both men and women.

MANY THINGS HAPPEN TO FOOD PRODUCTS AFTER THEY LEAVE THE FARM

Many changes take place in raw food products on their route

from the farm, through many channels, to the retail stores and then
to you.

> For instance, over five million men and women
> process our vegetables, pack our meat, bake the bread,
> and do all the other processing and distributing chores.
> Without them, our orange juice would never leave the
> grove, our steak would still be roaming the range.
> Trains and trucks and planes and barges are making
> longer trips from the farm to our bigger and bigger cities.
> Transportation gives us the benefits of year-round pro-
> duction in far away, specialized farm areas.
> We're getting more convenience with our food
> money—more trimmed, packaged, and frozen foods.
> More built-in maid service, such as pre-mixed foods and
> heat-and-serve dinners.
> All together, processing, transporting, and distribut-
> ing food take 65 cents of our food dollars.[1]

All this takes workers. And it also requires professionally
trained business specialists for supervision, management, planning, cost control, quality control, marketing policies, market surveys, finance, credit, purchasing, advertising, public relations, and
other professional services. Such requirements mean opportunities for college graduates.

HOW BIG IS OUR FOOD SUPPLY?

Each year the food industry assembles, processes, and sells
the food crops raised on a large portion of our 300 million harvested acres. And it processes, transports, and distributes about 33
million pounds of red meat (beef, pork, veal, lamb, mutton) and 7
million pounds of ready-to-cook poultry meat. Add to this 115 billion pounds (50 billion quarts) of milk processed into dairy products, plus countless other food specialties. You will be amazed at
the size of this vital industry. No wonder our larger retail stores
can carry 8,000 or more items on their shelves—mostly processed
foods.

---

[1]U.S. Department of Agriculture.

What Is Your Share of Our Nation's Food Supply?

You can get your answer from this table on per capita consumption for a recent year.

Food Consumption per Capita in Pounds

Dairy products (less butter) (fluid whole milk equivalent)	341
Eggs	36
Meats, fish, poultry	231
Dry beans, peas, nuts	17
Fats and oils (less butter)	52
Fruits and melons	154
Potatoes and sweet potatoes	101
Other vegetables	207
Flour and cereal products	142
Sugar and syrups	124
Coffee, tea, cocoa	13
Total	1,418

HOW MUCH DO WE SPEND FOR FOOD?

About $150 billion per year. Yet food is a bargain when compared with many other purchases. It has increased in price much less since 1947–49 than most other items.

> The American consumer will probably spend less than 20 percent of his annual disposable income for the best quality and greatest quantity of food available anywhere in the world. This is a considerably smaller proportion of our income needed for food than the Japanese who spend 42 percent, the West Germans who spend 45 percent, and the Russians who must use 56 percent of their disposable income.
>
> These countries also use much more of the low price cereal products in their diets than the expensive high protein sources used in America.[2]

Our farmers are efficient producers of raw materials. Our industry managers, administrators, and executives give us great vari-

---

[2]*Minnesota Marketing Messenger.*

ety and an abundance of high quality food at reasonable cost. But the industry needs new "blood" to maintain its forward progress. That's why its representatives come to the colleges to recruit young men and young women for responsible positions.

NOW LET'S LOOK AT SOME MAJOR PARTS OF THE FOOD INDUSTRY

First come the *producers of primary food products.* They are the farmers to whom we "tipped our hats" in Chapter 3. And they fully deserve our deep appreciation.

Next are the *manufacturers and processors.* We have more than 41,000 food processors and manufacturers in our nation.

Then come the *distributors.* They help get food products to you; they form a chain between the processor and your local store. This chain has many links—brokers, commission persons, purchasing agents, selling agents, and warehouse persons.

Finally there are the *retailers.* We have well over a quarter of a million—corner groceries, roadside markets, independent chain stores, giant supermarkets, discount houses.

A CLOSER LOOK AT FOOD MANUFACTURERS AND PROCESSORS

What They Do

Look over the shelves of a modern supermarket and you will see that a relatively small proportion of the thousands of items are fresh foods. Most of them are processed foods—canned, waxed, dried, frozen, bottled, pickled, packaged, wrapped, baked, changed in one way or another. Such treatments stabilize foods and keep them from spoiling. In fact, stabilization is a main purpose of processing.

> . . . to meet the food requirements of a constantly growing population, a technical revolution has taken place with respect to methods of handling, preserving, and distributing food. Not only have the older techniques such as drying, salting, canning, and preserving been improved, but new processes of dehydration, con-

FIGURE 5-1. Processing plants produce thousands of food items. (Courtesy, National Association of Retail Grocers)

> centration, quick-freezing, pre-cooking, and handling fresh fruits and products have been developed, together with related improvements in transportation, refrigeration, grading, packing, and sanitation.[3]

Each year you eat about 142 pounds of flour and cereal products that come from our great flour milling and baking industries, and you use 124 pounds of sugar and syrup, much of it coming from the confectioners. Indeed, products of our farms come to us as food in many different forms.

Some Manufacturers Are Small—
Some Are Large Corporations

Manufacturers and processors vary greatly in size and in annual business volume. You would describe some as rural industries, such as the potato chip manufacturers located in the center

---

[3]John H. Davis and Roy A. Solberg, *The Concept of Agribusiness*, Alpine Press, Inc., Boston, Mass., 1957.

FIGURE 5-2. Supermarket shelves are filled with processed foods. (Courtesy, National Association of Retail Grocers)

of a few hundred acres of potatoes. Actually, small manufacturers outnumber large ones. Such small manufacturers and processors sell their output within their own, or nearby, states. Thus, they provide good opportunities for business graduates who want to stay near home and grow into the management of a local concern. If you judge food processing by numbers and average size of firms, you would conclude that it is predominately a "small business."

On the other hand, you will find many large corporations with their stocks listed on the "Big Board" of the New York Stock Exchange. Some have shareholders in dozens of countries throughout the world. You often hear and read about them—concerns such as Bordens, General Foods, National Biscuit, Pillsbury, Charles Pfizer, Quaker Oats, Standard Brands, Kraftco.

Now We Get "Convenience" Foods

Processors and manufacturers give us convenience and other important advantages—high quality, great variety and abundance of items, all-season availability, good packages in various sizes and with visibility, standard grades, competing brands, and in many cases lower costs for a processed item than for its home-prepared counterpart.

An extension nutritionist at Colorado State University reports that homemakers now average about 1½ hours a day in preparing the day's meals; 20 years ago they averaged 5½ hours a day. The difference is due to the convenience of processed foods and the great variety now available—somewhere near 8,000 items.

Processing Plants Are Busy Places

They perform many operations. Before their products reach you, they may have been cooled, cleaned, graded, canned, frozen, shelled, sliced, mixed, and/or packaged, and some have been pre-cooked.

Sanitation and Cleanliness: When you visit a processing plant you'll be impressed with its clean look. Some of the first operations are washing and cleaning the produce, grading, sorting, removing waste portions, and preparing the food for table use. These and all other operations must have a completely sanitary environment. If they do not, the product might spoil, inspectors would be there with their "big sticks," and the business would soon decrease.

Operations: Here are some processing plant operations:

Refrigeration and temperature control may mean either warming or cooling. Certain fresh fruits and vegetables are pre-cooled right after harvest and then refrigerated while in transit; they are under controlled temperature and humidity during processing and while on their way to the cold show cases in the retail stores.

Freezing is another operation. Production of frozen foods increased about 4½ times during a recent decade. This not only has supplied our stores with frozen goods but also has given rise to the frozen food locker and freezer-provisioning industry.

Freeze-drying may well go down as one of the milestones in the history of food preservation according to a report from The

Ohio State University. Foods are first quick-frozen and then placed in a freeze-dryer subject to vacuum. The ice in the frozen products changes to vapor without even becoming a liquid. When the products are put in airtight containers, they can be kept for indefinite periods without sterilization or refrigeration, thus reducing the cost of storage and handling. Food is rehydrated quickly and thoroughly by placing it in water for recommended periods. Loss of flavor is minimized.

Hydrocooling involves passing the product through icy water or immersing it in water. The process is commonly used for peaches, sweet corn, celery, asparagus, and other vegetables.

Vacuum cooling is a fast method for reducing temperature that is adaptable to large scale operation. One California plant, for instance, has a cooling capacity of 75 carloads a day; some plants have vacuum tanks large enough to take in a railroad car.

Dehydration methods are being constantly improved so that practically instant dehydration of vegetables and other foods is now possible. Most procedures are economical and relatively easy to perform. High-vacuum process removes almost all the water, so the products store well.

Irradiation, or sterilization by electron radiation, although not yet commonly used, shows promise for the future. The Quartermaster Corps of our Army and the Atomic Energy Commission have led in the research and development of this new method of food preservation. Irradiation does not make the food radioactive.

Additional Important Processes: Precise, automatic control of temperature, humidity, environment, and timing is essential in the many and varying operations such as these:

Canning is a time-honored method giving convenience, storage at room temperature, and safe products. Recently improved methods (*e.g.,* the high-short process) employ high temperature heating for short periods.

Concentrating involves a vacuum process which, at reduced temperatures, quickly removes the water from soups, milk, fruit juices, and similar products.

Pasteurizing is a method of milk heating which destroys all or most micro-organisms and helps retain flavor and color.

Blanching involves cooking pieces of food a short time in

water or steam to inactivate enzymes and afford better retention of color and flavoring during storage.

Sulfiting is exposing the food product to sulfur dioxide gas or dipping it in a liquid containing a similar substance. This, like blanching, protects color and flavor.

Mixing is another important operation. Dry bakery powder mixes for cakes, pie crust, rolls, cookies, and muffins have become quite popular and substantially reduce time needed for home preparation of these foods.

Adding chemicals to foods is not a new process. We use a great many chemicals, as many as 700. They add color, prevent mold, reduce spoilage, enrich the food, and make it more nutritious.

Under present laws, a manufacturer of a food chemical must prove its safety before it can be sold for use in foods.

Packaging and Packing: There are critical operations in food processing, so important, in fact, that a separate industry has developed to supply packaging materials. Now we use many different materials to replace the older food containers such as the cracker and pickle barrels the grocer used to dip into when he doled out the customer's portion.

Suppliers of packaging materials include many companies; some are small, and some are large. Continental Can, for instance, has 140 plants at 94 cities throughout our nation supplying materials of plastic, paper, foil, film, glass, and metal. Other firms supply shipping cases, pallets, boxes, cardboard, and a profusion of other materials.

You can measure the importance of packaging by our expenditures for it:

"Of every $20.00 we spend for groceries, from $1.50 to $2.00 is for packaging," reports Virginia Polytechnic Institute. Our total costs for packaging run over $30 billion per year. You can see why packaging has become an integral part of the product:

> It makes possible self-service in the store.
> It allows consumers to see what they buy.
> It protects against contamination—preserves the product.
> It gives us unit-size packages—with number of servings noted.

It helps gain consumer acceptance—sells the product.
In many cases, it lowers the cost of retailing and lessens
    waste.

Try to Visit a Processing Plant

As you observe the plant operations, you probably will consider most of these as packaging activities: trimming, washing,
pre-weighing, counting, chapping, filling, closing, sealing, convey-
ing. Automatic machines with precise electronic controls govern
many of these operations. New machines and more efficient
equipment are always sought; they hold down production costs
and produce even more attractive products.

You would meet "specialists" like these in a processing plant:

> Warehouse and shipping manager
> Processing equipment engineer
> Equipment service supervisor
> Quality control supervisor
> Processed food salesperson
> Plant department supervisor
> Production control supervisor
> Fieldperson
> Sanitarian
> USDA grader
> Food inspector
> Cold storage manager
> Produce buyer
> New product developer

Processors Constantly Seek New Products

Rhode Island Resources (formerly *Rhode Island Agriculture*),
a publication of the University of Rhode Island, calls our attention
to the many new products striving for public favor:

> In the final analysis, it is what the public likes that
> stays on the supermarket shelves. These stores carry
> from 6,000 to 8,000 products and often even more. There
> is no room for a product that doesn't sell. Shelf space is
> valuable and there are a host of even newer products
> that are clamoring for attention. Food buyers are offered
> 24 new products a day, 120 a week, or more than 6,000
> every year. Only about one in ten new products lasts as

long as a year, but even at that, between 60 and 70 percent of the items on the grocers' shelves were not there 10 years ago.

Food Processors Have Their Problems

Like all progressive manufacturers, food processors look forward, not backward, although they do learn from yesterday's mistakes. They are always seeking improvements, lower manufacturing costs, wider markets, new products.

Some of their problems are technical; some are managerial or administrative. Others relate to finance, budget, and advertising, and others to public relations. Here's one you can think about right now.

The president of one agribusiness corporation, in discussing

FIGURE 5-3. This senior student is being interviewed for a position in the chemical food industry. (Courtesy, University of Arizona)

"what presidents think about at night," said no other single element was more important in the success of his business than "how his people get along with each other."

Test yourself on that one. How do you get along with your associates and your classmates?

You can see that running a business or helping to solve its problems brings many challenges. It calls for versatility, energy, courage, and judgment. Look, for instance, at this challenging five-point program of a leading canning company:

1. Tight cost control
2. Plant modernization
3. Two-phased research and development program
4. Integrated long-range planning
5. Product diversification

Looking for new products and new methods of processing requires continual research. It is costly, but it is the key to progress. Food processing firms spend more than $150 million annually for basic and applied research.

GOOD OPPORTUNITIES IN ALL PARTS
OF THE FOOD INDUSTRY

A large food business, the Carnation Company, has this to say to potential employees:

> When we hire you we make a commitment to train you just as fast as you can move ahead. You won't find yourself sidetracked, because we don't overhire, and our business continues to grow. Each year new products and additions to our established product lines create new jobs on all levels.
>
> If exciting prospects for growth in a secure industry are what you are looking for, get in touch with us. Let's explore opportunities in sales, production, product management, research, and dozens of staff departments. Geographically, we have plants and offices from one end of the country to the other.

Many other companies dealing in food can offer you similar opportunities.

The Wilson Food Corporation suggests that you select a possible position, and then take a course of study to prepare you for that position. Wilson's program is as follows:

	Accounting	Animal Science	Business Admin.	Chemistry	Engineering	Economics	Liberal Arts
				Courses of Study			
Accountant	X		X			X	
Maintenance Engineer					X		
Industrial Engineer					X		
Livestock Buyer		X					
Product Department Manager	X	X	X			X	X
Production Supervisor		X	X	X	X	X	X
Quality Control Supervisor					X		
Research & Development Supervisor					X		
Salesperson	X	X				X	X

Now let's examine specific work areas and positions in these main parts of the industry—processing and manufacturing, distributing, and retailing.

FIGURE 5-4. Research is necessary to obtain new and better varieties. (Courtesy, Agway Inc.)

Opportunities in Processing
and Manufacturing

Your college training in agribusiness or in majors relating to it will give you excellent fundamental preparation for many kinds of employment.

1. *Purchasing Raw Materials:* If you have a farm background, perhaps you will have an advantage in this activity. A job may require contacts with farmers, cattle feeders, and fruit

FIGURE 5-5. View of fresh meat processing plant. Manager can see some of the major processing areas from his desk. (Courtesy, Agricultural Marketing Service)

and vegetable growers and may require knowledge of farm products. Knowledge of the problems and customs of the producers is desirable as well as an understanding of costs and prices. Processors frequently contract to take a whole year's production from groups of farmers.

Purchasing may require contacts with wholesalers, brokers, produce merchandisers, and others. It will often involve travel and outdoor activity. Ability to meet people easily and to get along well with them would be a real asset.

2. *Plant Managers and Supervisors:* Business training, plus product knowledge, enables the manager to keep production equipment busy. Planning production schedules to avoid shut-downs and slow-downs is essential to economical operation. Your company can't make a profit from unused capacity. Flow-charts, sales records, statistics, market surveys—all help the manager map out a program that will keep the technicians and operators busy turning out optimum production for every season.

Good managers develop these abilities, among others:

To see and evaluate new opportunities
To measure—objectively—the progress of their
    operations
To plan ahead
To organize operations
To develop employees' enthusiasm

3. *Inspectors for Quality Control:* These specialists seek high quality at all stages—at the source of supply, when the raw product arrives at the plant, when the finished product leaves the plant, when it is being transported, and when it is in the retail store.

High quality and good appearance affect sales fully as much as price. Research scientists have developed instruments to measure quality in food products—the amount of fat in meat, the maturity of apples, etc. Colorimeters measure color in various products, and instruments check and reveal defects in a variety of food products and determine chemical and nutrient constituents.

Desirable qualifications for inspectors include knowl-

FIGURE 5-6. Here are some books you may use in your food courses. (Courtesy, S. Pendrak, SUNY–Cobleskill)

edge of farm products: where they are grown, conditions under which they are grown, how they withstand shipment or storage, what the refrigeration requirements, standards, grades, and consumer desires are.

Most of the inspectors' work may be within the processing plant, but they too may make field trips to visit producers or suppliers.

4. *Salespersons:* Every industry depends on sales. That's why there are so many openings in this field. Competition is keen among food processors. Their customers—the food buyers—have a wide choice of suppliers and products, especially new products.

Agribusiness graduates have special advantages as salespersons. They have a thorough knowledge of their products and have studied fundamentals of salesmanship—sales promotion, organization, market surveys, advertising, and communications. And, if you choose sales, the processing company will give you excellent training in its in-service sales school. There you will meet and be guided by experienced salespersons.

Good salespersons know the community they serve. They study trends, employment, population changes, and new industries entering their territory. They aid the dealer or wholesaler in many ways, so relations with customers are enhanced.

Successful salespersons are in a favorable position. Not only do they make good incomes, but also the sales department is very often a stepping stone to an executive position of greater responsibility.

5. *Cost Control:* Where do losses occur? What departments or operations have excessive losses? Are losses due to in-plant operations, careless purchasing, spoilage, costly shipping methods, inefficient marketing? All these items and others bring challenge to the business-trained specialist. Your courses in accounting and the overall view you get from courses in economics will be of great assistance to you here. Cost control requires careful and exact attention to detail and keen analysis.

6. *Pricing and Contracting:* What's the right price for a particular item? Do you know exactly what the item costs, how
much overhead expense should be charged against it? Will
your proposed price meet competition? Will the item sell
at your price?

Who is the best supplier? Should you make a long-
term contract? What quantities should you contract for?

In this area you will find challenges and opportunities
for important professional service essential to the success
of the whole processing operation.

7. *Finance, Capital, and Budget:* Where can the manufacturer
most profitably invest money? What product will be most
profitable? How much expenditure will the year's operation require? How can the manufacturer raise the necessary working capital? Does the company have too large a
proportion of receivables? All these and many more challenges confront the financial managers.

In a large concern, the functions related to these questions may be placed in the treasury department. That's a
good place to start your career if finance is your chief interest. Of course, you will not start at the top, but you may
have the privilege of associating with seasoned financial
executives. You can gain valuable experience, and you'll
be on the road to advancement.

Your courses in economics, banking, money, credit,
corporation finance, investments, and accounting will help
prepare you for work in this area.

8. *Transportation and Shipping:* Agribusiness graduates have
advantages here also. They know that farm products are
relatively bulky and that many are perishable. In many
cases, the products are shipped long distances. Today, not
many items are ever "out-of-season." Some areas of our nation are almost always producing them.

Accordingly, food products are quite sensitive to transportation costs. For instance, shipping a head of lettuce
from California to Boston adds about 10 to 12 cents to its
selling price.

Transportation specialists constantly strive to reduce

costs and to improve services by shippers, carriers, and receivers. They make field trips; visit packers, shippers, and handlers; study state and national transportation policies; gather cost data; test various methods of shipment; and experiment with various routes.

Statistics, records, costs, reports, complaints, studies of delays, information about alternate routes, and similar data are employed in this special function.

9. *Communications, Advertising, Public Relations:* The food industry is so close to the public that it must constantly strive to maintain a good public image. We are quick to be critical when anything goes wrong. Adverse criticism must be offset with constructive good-will-building efforts.

All media are used—radio, TV, newspapers, magazines, poster displays. Some efforts are for national coverage, some for local and community interests.

Graduates trained in advertising and communications who demonstrate ability and determination can find many opportunities to work for manufacturers and their trade associations.

Competition between different brands of foods and the great effort necessary to introduce new brands or new food items call for many professionally trained workers.

10. *Positions with Related Industries:* Many other manufacturers and suppliers serve food processing companies. The things they make and the services they provide are closely related to food products. These related industries need trained employees who know agriculture, farm products, and business practices.

Here are some examples of related areas:

Packaging
Canning
Warehousing
Cold storage
Materials-handling equipment
Processing machinery and equipment
Instruments and controls
Packaging film and cartons
Aromatics and flavors
Food additives

Opportunities in Wholesale Distribution

Wholesalers help get food products to you. We could over-simplify and say the distribution chain is from grower—to shipper—to carrier—to wholesaler—to retailer—to you. But there's more to it than that. Other methods are used also. Of course, distribution does depend on wholesalers; 25,000 of them handle our fresh produce. And distribution depends on transportation and on retail outlets. But it also includes brokers, commission persons, purchasing and selling agents, auctioneers, and persons involved in warehousing and storage.

Distribution provides employment opportunities with:

> Wholesalers
> Brokers
> Commission persons
> Public auctioneers
> Purchasing agencies
> Selling agencies
> Fruit and vegetable packers
> Warehouse and storage services

1. *Wholesalers:* The major function of wholesalers is to distribute to other firms (usually to retail firms) within their territory. This area must be large enough to warrant keeping large stocks at a central point. Usually wholesaling is a relatively large venture, requiring considerable capital. "Merchant wholesalers" buy and sell primarily for their own accounts. "Service wholesalers" perform other functions also, such as warehousing, delivering, extending credit, providing market news, sending out salespersons, and assisting their retail customers with merchandising.

   Some wholesale operations are big business. Take Super Valu Stores, Inc., for instance, a Minnesota firm that services over a thousand of its own retail stores.

   You will find that distribution is a rapid-action process. It has many parts and requires many special services. More than you think! A recent "work-shop" conference of distributors at the University of Delaware covered 20 different phases of distribution.

2. *Supervisors of Fruit and Vegetable Packers:* Packers of fresh fruits and vegetables often serve right where the items are grown. In fact, packing agencies may be owned by growers' associations. Supervisors must know the products; they must be able to maintain careful grading and labeling and expert packing. They must be alert and receptive to new methods of reducing packing costs and avoiding wastage.

3. *Purchasing Agents (Buyers):* Much of their selecting and purchasing is done at the shipping point and requires contact with growers and packers.

4. *Employment with Brokers:* Brokers buy at shipping point. Often a broker sends the product to a distant buyer. The buyer can safely purchase from a distant source because of rigid standards and grades and careful inspection. Brokers usually do not take title to the merchandise; their principal function is to bring buyer and seller together.

5. *Employment with Shippers:* Shippers usually buy directly from growers or associations and ship in large quantities to established customers. They perform a service-type function.

They employ college-trained people who can start

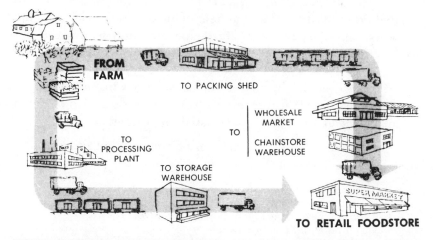

FIGURE 5-7. Principal marketing channels for fruits and vegetables. (Courtesy, United Fresh Fruit and Vegetable Association)

with a basic knowledge of the products and an understanding of business fundamentals and transportation problems.

With fresh fruit and vegetable products, distribution requires fast action and often quick diverting of a shipment from its original destination to another point where demand is more urgent.

One shipper-broker, who handles cherries, grapes, and some other fruits, has an excellent in-service training program for new employees. He employs about five agricultural graduates each year, then gives them a chance to travel and see his whole operation. His new employees visit growers, packers, and warehouses in the western states; study transportation and shipping problems; and then visit some major customers in the midwestern and eastern states. This shipper is training future managers. He wants them to know the business "from the ground up." He also works with colleges in developing curricula.

6. *Service with Auctioneers:* Auctioneers sell fruits and vegetables at shipping points near the growers—or in large terminal markets. You would see no produce in a terminal market auction. The buyers consult lists of items previously inspected and graded to rigid standards.

Business-Agriculture Education Prepares for Careers in Food Distribution: Within the channels of distribution, you will find packers, shippers, brokers, storage persons and warehouse persons, commission persons, merchants, auctioneers, and wholesalers.

It's an exacting business. Mistakes, errors in judgment, and faulty decisions can cause serious losses. But it can also be profitable and mean a fascinating and satisfying career.

Summary: A quick summation of employment opportunities in wholesale food distribution would include:

> *Buying*—including inspection, packing, grading, labeling, and pricing.
> *Storage, warehousing,* and *quality control.*
> *Shipping and transportation.*
> *Selling*—pleasing customers, obtaining new customers, maintaining adequate selling prices.

Opportunities in Retailing

The Independent Grocers' Alliance (IGA) has this to say about a career in the retail food field:

> Food retailing makes some strong demands upon its employees. If you are in doubt about meeting those demands, your future may not be in the food business.
>
> Lacking the proper temperament, attitude, and ability to work with others, some persons might be better suited to desk-bound jobs where they can work alone.
>
> What about you? Do you have what it takes? Read the questions that follow and find out. If you can answer "Yes" to them, you could have a bright future in retailing.
>
> 1. Do you like people? Do you like to work with others, to work as a team, to deal with all kinds of customers?
> 2. Do you like food retailing? Having read about this business, does it sound appealing, fascinating? Does merchandising offer a challenge to you?
> 3. Are you service minded? Do you enjoy pleasing people, hearing their complaints, and helping them? Could you give service with a smile at all times, no matter how you feel?
> 4. Can you work under pressure? At peak periods when customers crowd the checklanes, can you work calm and collected and efficiently? Can you take emergencies in stride?
> 5. Your character: Is it good? Are you honest, and do you believe in a day's work for a day's pay? Can you be depended upon?
> 6. Are you ambitious? Do you want to learn all you can about the business by working, reading trade journals, undergoing training programs? Do you want to get ahead but realize it isn't an overnight jump?

Retail sales is the final link in the chain of distribution from the farm to the dinner table. Look around and you will be amazed at the multitude of retail establishments—big and little. You'll notice everything from the roadside drive-in market to the giant supermarket and the discount house—chains, independents, cooperatives, farmers' markets, delicatessens, automatic vending machines; never before have you had such a wide choice of places to buy.

Whatever the type of establishment, it must have good management to stay in business. Retailers stake their future on their ability to have the desired items on hand in good condition and well displayed in order to maintain their customers' good will and continuing patronage.

If you included every type of retail food outlet, you would find over a quarter million enterprises. But most of these are quite small. Although often the source of pride and good profit to their owner-proprietors, they are responsible for only a minute portion of our total food purchases.

Many Employment Opportunities in Retailing: Retailing has become a complex business. It requires many "specialists." That's why this list of work opportunities is so long:

> Store manager
> Buyer of food products
> Buyer of store equipment
> Fieldperson to contract with farmers
> Produce specialist
> Meat specialist
> Dairy product specialist
> Sales promotion manager
> Appraiser of new products
> Storage and warehouse manager
> Traffic and shipment specialist
> Market survey researcher
> Advertising and public relations personnel
> Customer relations specialist
> Consumer acceptance researcher (determines who likes
>   what)
> Store layout and in-store traffic specialist
> Owner-proprietor

Formal education in business and agriculture will enable you to enter any one of the foregoing areas of service.

Retail Stores Growing Larger: Like many movements today, the trend in food retailing is toward big business, mergers, and consolidations. This trend has brought us the one-stop shopping center, self-service, all-season product availability, professional packaging, standard grades, and carefully inspected, well preserved products.

FOOD STORE ORGANIZATION CHART

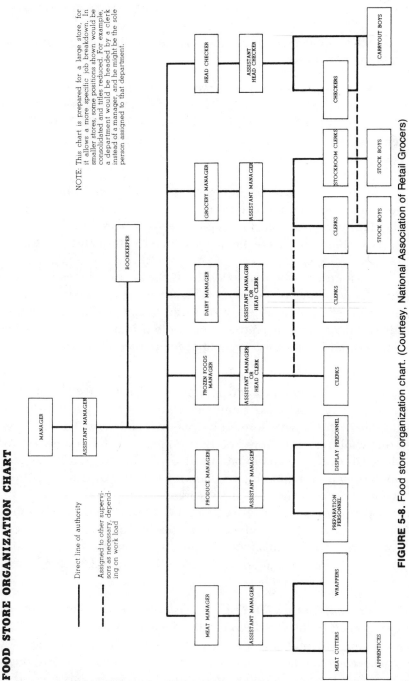

Direct line of authority

Assigned to other supervisors as necessary, depending on work load

NOTE: This chart is prepared for a large store, for it allows a more specific job breakdown. In smaller stores, some positions shown would be consolidated and titles reduced. For example, a department would be headed by a clerk instead of a manager, and he might be the sole person assigned to that department.

FIGURE 5-8. Food store organization chart. (Courtesy, National Association of Retail Grocers)

So let's look at the organizations that do the major part of the retail food business.

Chain Stores and Supermarkets: The U.S. Department of Agriculture estimates that chain stores and supermarkets do 70 percent of the retail food business. They carry from 10,000 to 12,000 items on their shelves; many of these come from distant sources. There are items from almost every state and from some foreign countries.

Even though we consumers have our likes and our dislikes, we accept new products quite well; at least we have in the past. About 70 percent of our food purchases—measured in dollars—go for food items that have been introduced within the past six or seven years.

A fast-growing type of foodstore is the "convenience store"; it offers convenience of location, quick service, and long store hours. Such features have enabled these stores to compete despite higher prices, larger margins, and a more limited brand selection. They now account for almost 5 percent of total grocery sales.

Independent Stores: Independent store operators and owners have merged, enlarged, and affiliated and have become participants in big business. Some have affiliated with wholesale suppliers; others have formed cooperative groups which may perform wholesale functions, or the groups may be sponsored by wholesalers. These combinations give smaller enterprises greater purchasing power and other economies so that they compete effectively with the chains and supermarkets.

Some independents enlarge by licensing or franchising other retail stores which are independently owned and operated. These stores are given management counsel, personnel training, and competent engineering advice in store lay-out and in purchasing facilities and equipment.

Discount Stores: During the last few years discount stores have spread to the retail food industry. When you visit a discount store you will probably find a well stocked food department. A typical discount house may carry some 20,000 items. Although more than half are non-food items, the variety of foods offered is

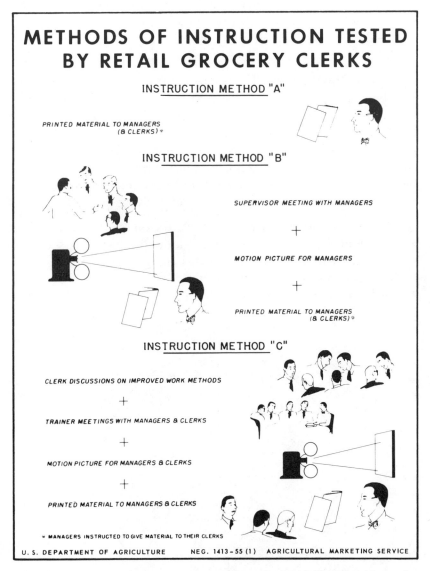

METHODS OF INSTRUCTION TESTED BY RETAIL GROCERY CLERKS

INSTRUCTION METHOD "A"

PRINTED MATERIAL TO MANAGERS
(8 CLERKS)*

INSTRUCTION METHOD "B"

SUPERVISOR MEETING WITH MANAGERS

+

MOTION PICTURE FOR MANAGERS

+

PRINTED MATERIAL TO MANAGERS
(8 CLERKS)*

INSTRUCTION METHOD "C"

CLERK DISCUSSIONS ON IMPROVED WORK METHODS

+

TRAINER MEETINGS WITH MANAGERS & CLERKS

+

MOTION PICTURE FOR MANAGERS & CLERKS

+

PRINTED MATERIAL TO MANAGERS & CLERKS

* MANAGERS INSTRUCTED TO GIVE MATERIAL TO THEIR CLERKS

U. S. DEPARTMENT OF AGRICULTURE   NEG. 1413-55 (1)   AGRICULTURAL MARKETING SERVICE

FIGURE 5-9. Seeking the best method for "communications." Methods of instruction tested by grocery clerks. (Courtesy, Agricultural Marketing Service)

tremendous. Today, the annual food sales of these stores exceed $2 billion. We may see many more discount food stores in the years ahead. This is an industry trend; even major supermarkets and chains are now planning to enter this field. It may bring one-stop shopping for both food and non-food items.

Cooperative Stores: These are controlled by share-holding members. Earnings are refunded to customer-owners in proportion to their purchases. Mergers are taking place among the co-ops also; they are becoming bigger and fewer. Larger, affiliated groups, merged together, make these enterprises more like the chain-store organizations. And they may bring economies—more buying advantages, lower transportation and delivery costs, fewer executives, larger volume, and better advertising programs.

The Food Industry Is a Growing Business: Retail sales increase about 5 percent a year. Look around and you'll see new stores in many new locations. A food chain now has some 40 stores in the Washington, D.C., metropolitan area; plans call for 19 more. A still larger supermarket chain has over 250 stores in its Washington division. Another chain near Minneapolis plans to add 48

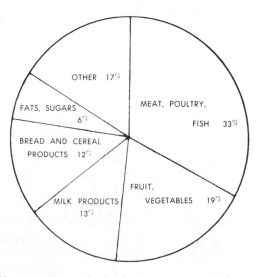

FIGURE 5-10. How we spent our food dollar in a recent year. (Courtesy, The Pennsylvania State University)

more stores in its 12-state area. New stores come because there will be more customers. That means increased activity in every part of our food processing and distributing system and also in food production.

OUR POPULATION INCREASES RAPIDLY

Stand in the lobby of the U.S. Department of Commerce in Washington and watch the "clock" that counts and records our nation's growth—a birth every 9 seconds, a death every 16½ seconds, an immigrant coming into our country every 60 seconds, an emigrant leaving our country every 23 minutes. All this gives us a net increase of one every 15½ seconds. Figure it out for 24 hours and you'll find that we have over 5,000 more mouths to feed every time a new day dawns.

THE FOOD INDUSTRY HAS PROBLEMS AND OPPORTUNITIES FOR YOU

Fast-growing and rapidly changing, all parts of this great industry have problems. That's why their representatives are seeking young men and women who have had formal training in business and agriculture.

You may find your career in the food industry.

Chapter 6

THE DAIRY INDUSTRY

THE DAIRY INDUSTRY

Another big business! The dairy industry buys 115 billion pounds—50 billion quarts—of milk from producers each year. From this huge amount it makes a multitude of dairy products; transports, distributes, and sells them everywhere in our country; and sends them to many foreign countries.

To do all this it employs some 187,000 workers, along with the necessary supervisors, managers, administrators, and specialists. They work in thousands of plants, distributing centers, retail stores, and offices throughout our land. Some dairy enterprises are large and national in scope; some are small and serve only their local areas.

If you computed the yearly retail value of all kinds of dairy products, you would need a long string of zeros; your total would be about $11.5 billion. These products supply close to a fourth of our total nutrient requirements. The dairy industry is one of our largest. It supplies a steady demand because of the importance of milk in our diets. Its future growth seems assured because of our increasing population. Milk is one of our cheapest foods in terms of nutritive value.

THE DAIRY INDUSTRY PERFORMS MANY OPERATIONS

Here are some of the activities the dairy industry performs to prepare its products and get them to you:

Cooling and Refrigeration: This may begin at the farm, where raw milk from the milking machine is piped into refrigerated bulk tanks. Temperature control is essential for most dairy products all the way from the farm to the consumer.

FIGURE 6-1. About 11 million dairy cows give us 50 billion quarts of milk each year.

Procurement and Assembly: Refrigerated tank trucks take the milk from the producer to the processing plant or receiving station. Much of it is purchased from the consumer under government regulated marketing orders, which set prices based on butter fat content and grade.

Processing and Manufacturing: These include pasteurizing and homogenizing fluid "whole" milk and manufacturing the raw milk into scores of dairy products.

Bottling and Packaging: These involve putting the milk or milk product in bottles or other containers that will preserve the product well and make it available in sizes that will be most convenient for the consumer.

Transportation and Delivery: These are particularly critical activities because of the nature of the products and the need for accurate temperature control. Probably more motor trucks—most of them refrigerated—are used in transporting and distributing milk and its products than in moving any other commodity.

Retail Sales and Delivery: Whenever you travel in the United States, you can always get clean, fresh, cool milk. What an amazing

FIGURE 6-2. Modern dairy farm building made of concrete, glass block, and redwood. It is neat, attractive, and serviceable. (Courtesy, Ohio Agricultural Research and Development Center, Wooster)

tribute this is to the dairy industry and to the technicians and scientists who serve it!

. And how convenient it is to have dairy products delivered to your home—day in and day out, rain or shine. Not only is the dairy industry a manufacturing industry but also a service industry, but fewer and fewer people now require home delivery.

When you think of these services and when you see the great variety of dairy products so attractively displayed—and cooled—in your favorite store, think of the vast amount of planning, management, and efficient administration that makes this great industry function so well.

Regulation, Inspection, Public Health: Milk and dairy products are probably more strictly regulated, inspected, and controlled than other farm products. Without such rigid controls they might become health hazards and disease carriers.

Federal, state, municipal, and other local officers carry on this

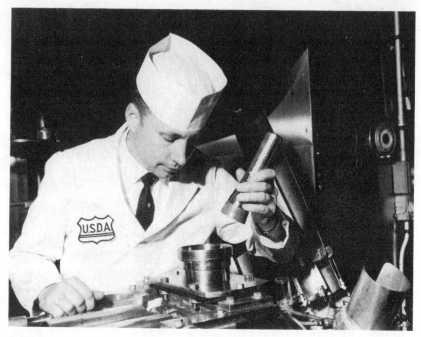

FIGURE 6-3. Inspector checking cottage cheese packaging machinery. (Courtesy, USDA)

health-guarding work. All segments of the industry—from producer to retailer—are subject to some form of inspection and regulation. About half of the milk sold by farmers is subject to federal marketing orders. Such orders—approved by vote of the producers—are designed to assure dairy farmers of a minimum price based on current supply and demand conditions. Such federal orders apply only to prices paid to farmers by milk dealers; wholesale and retail milk prices are not subject to federal orders but are set by competition, unless controlled by state laws.

LOOK AT SOME OF THE MANY OPPORTUNITIES IN THE DAIRY INDUSTRY

It's a big business, and practically everyone is its customer. No matter what your special work in it may be, your chance for advancement will depend largely upon having a solid foundation—a knowledge of sales, merchandising, management, traffic control, cost control, and accounting, as well as of the basic technology of the industry. You should have formal education in both business practices and dairy industry problems.

DAIRY PROCESSING OR MANUFACTURING

This fundamental part of the tremendous dairy industry is well described by Michigan State University:

> What is the dairy processing industry? It is that branch of the food industry which deals with the procurement, processing, packaging, and marketing of 115,000,000,000 pounds of milk produced annually in the United States and the products made from it. The major products, processed or manufactured, include bottled milk, butter, cheese, ice cream, yogurt, condensed and evaporated milk, and dried milk. Sometimes this branch of dairying is referred to as "Dairy Manufacturing" or the "Dairy Manufacturing Industry."

Specific Job Opportunities in Dairy Processing

Field Service: Fieldpersons work with the dairy farmers to assure the processing plant good sources of fresh milk. Also they may help milk producers improve their herds and their methods. Graduates with farm backgrounds or those who wish to work near home may welcome positions as field workers. Banks and breed associations also employ them to work with dairy farmers. Field work is good experience—a good starting point that prepares you for greater responsibilities.

Procurement and Purchasing: Although fieldpersons procure raw dairy materials, other specialists also may perform this function, in addition to purchasing supplies and equipment for plant operation.

Plant Operation: In-plant operations afford many jobs for specialists—operators, supervisors, process managers, and maintenance managers for the many intricate, automatic machines.

Quality Control: Dairy products are subjected to many in-plant tests and inspections, which are performed by specially trained inspectors and graders. Without this essential service, products would fail to meet the rigid state and federal standards.

Sales Managers: Profitable manufacturing depends largely on volume of sales. Idle plant capacity usually means loss. Efficient, active sales and merchandising are vital to this highly competitive business. If your plant doesn't get the business, others will. Graduates who have, or who can develop, modern merchandising methods can find a rewarding career in this field. They should be well grounded in advertising, sales promotion methods, communications, and public relations.

In-plant Business Services: These include opportunities for graduates who want careers in finance and budget control, insurance, accounting, advertising, personnel work, labor relations, public relations.

Managers and Executives: These are top positions. Perhaps

FIGURE 6-4. Cooperative milk processing plant. (Courtesy, Dairylea, Inc.)

they are your goal. To reach the top, you will need some years of experience and an outstanding record of achievement on the steps that lead to the top. You can make a good start while in college through a combination of courses in business and dairy manufacturing.

A *Business of Your Own:* Big corporations account for the largest share of manufacturing activities. But because the dairy industry includes so many products, it does afford a good chance for the success of relatively small operations. Usually a small concern concentrates on a few closely related products—perhaps only one—ice cream, for example. The local plant operator has certain advantages—acquaintance and community interest and support. Several dairy manufacturing graduates have succeeded in conducting their own processing business. Owning dairy manufacturing businesses has special appeal for those who want to return to their home communities.

SELLING IN THE DAIRY INDUSTRY

Selling Dairy Plant Equipment

This is a special field in itself, a service for producers, processors, and distributors.

Look over a modern dairy farm and you will see a highly mechanized, sanitary, smooth-running operation. You will see milk conveyed through clean pipe lines to the refrigerated bulk tank, where it is automatically weighed and recorded, and you will see accurate control and proportioning of each cow's ration. The pieces of equipment that accomplish these tasks are not "over-the-counter" items. They are procured and installed after consultation and careful planning with the manufacturer's sales personnel who have real knowledge of the problems. And when you see the great variety of intricate mechanical and electronic installations used by processors, you will realize that dairy manufacturing is highly mechanized and automated. Processors are constantly seeking to lower the cost of plant operations and to make these operations sanitary, rapid, and automatic. That's why you will see so much rapid-action equipment—bottle fillers and cappers, package

FIGURE 6-5. Food store manager explaining to a foods class the problems of a retail store operation. (Courtesy, Cornell University)

fillers and sealers, electronically controlled conveyors, loaders, pasteurizers, vacuum pumps, temperature controls, and scores of servo-mechanisms.

Take the distributor who has specially designed delivery trucks. Even though the dairy industry uses more trucks than are used in transporting any other single product, designers and dairy specialists are still striving for improvement; and improvement will come only through the efforts of those who know the business "from the ground up."

Selling Dairy Supplies

Selling dairy supplies to producers and processors also calls for trained specialists. Formula feeds, with their health-protecting additives, are best selected and supplied by persons with knowledge of the sciences affecting nutrition. Likewise, supplies and chemicals for processors are critical items that must be carefully prescribed.

Salespersons of dairy equipment and supplies combine consultant service with selling; in a real sense they render a professional service in a highly specialized field. This constitutes a promising career for a graduate of a combined business and dairy manufacturing curriculum.

PREPARING FOR CAREERS IN THE DAIRY INDUSTRY

Because this industry is so important and so active everywhere in our nation, most of our colleges of agriculture offer several ways of preparing for careers in it.

Curricula combining dairy manufacturing and business administration are offered in most states. The University of Georgia offers the following options:

Dairy Manufacturing with Business Administration Option: This program affords major courses in dairy manufacturing and a maximum of courses in business and economics. If your major interest is in the business aspects of the dairy industry, you would do well to choose this combination.

Business Administration and Dairy Plant Management Option: This program is administered by Georgia's College of Business Administration. It prepares one for a career in general business, sales, personnel work, and accounting.

Sanitary Science: This is a special program and curriculum to prepare one for work in the field of public health and for work of professional status connected with the many regulatory functions that are closely related to the dairy industry.

Dairy manufacturing and its related activities make up a broad field. The many subjects in a typical curriculum give a well rounded business, technical, and cultural training.

Look at this condensed list of dairy manufacturing subjects offered at *Kansas State University:* market milk and dairy inspection; manufacture of ice cream, powdered milk, butter, cheese; elements of dairying; fundamentals of dairy technology; technical control of dairy products; judging dairy products. Also included

FIGURE 6-6. Class interviewing a dairy farmer on his problems in producing milk on the farm. (Courtesy, Cornell University)

are microbiology, bacteriology, general chemistry, organic chemistry, biology, economics, oral and written communications, general psychology, psychology of advertising and selling, commercial correspondence. And you can choose still other business subjects: cost accounting, sales management, retailing, business organization and finance, marketing, labor economics, business law, money and banking, credits and collections, marketing dairy products, principles of cooperation.

Such a wide choice of subjects leading to a Bachelor of Science degree gives you a splendid opportunity to get a good general education and, at the same time, prepare for a career in the phase of the dairy industry that seems most attractive to you.

What Do Processors Want in College Graduates?

Plant field work requires imagination, analytical thinking, and planning. Processors seek graduates with three characteristics:

namely, (1) an open mind, (2) enthusiasm, and (3) a sense of responsibility.

Graduate Study

If you can possibly do so, it will pay you well—in many ways—to continue your education and earn a master's or even a doctor's degree. Highly trained specialists and research scientists are urgently needed in the dairy industry, as well as in other industries founded on the sciences. And what industry is not dependent on science?

There are great opportunities provided by the need for still greater improvement in methods of production, processing, preservation, and distribution of dairy products. The steady progress of the industry thus far has been due in large measure to the research and investigations of scientists who had specialized training at the post-graduate level.

CONSIDER THE DAIRY INDUSTRY IN YOUR CAREER PLANNING

It has many opportunities for you, whether you want to work near home or far afield. It is not overcrowded with college-trained personnel. It is one of the most important agribusiness fields.

Consider the following questions: What positions in the dairy industry appeal to you the most? Do they provide good opportunities for advancement? Should you get an advanced degree?

Chapter 7

THE GRAIN INDUSTRY

THE GRAIN INDUSTRY

When we considered "goods from the earth" in an earlier chapter, we might have given special mention to the billions of bushels of grain our farmers produce each year. These huge amounts make up a major part of the "merchandise" of agribusiness.

THE GRAIN INDUSTRY MOVES MOUNTAINS

Now look at the grain industry, *an agribusiness that moves mountains*—mountains of food grains, feed grains, and oil seed grains. The grain industry moves these mountains from the farmer-producers to the processors, manufacturers, flour millers, feed millers, livestock and poultry feeders, dairy workers, brewers, distillers, exporters, oil seed crushers, and other grain users.

This year-round moving service takes lots of specialists who know the exacting details of their particular phase of the business. It takes shippers, brokers, commission persons, elevator operators, warehouse persons, transportation experts, insurance persons, bankers, traders in futures, and cash grain buyers and sellers. So there are many opportunities for you.

GRAINS ⟶ PROCESSORS ⟶ MARKETERS ⟶
MANUFACTURERS ⟶ CONSUMERS

SPECIFIC POSITIONS IN THE GRAIN INDUSTRY

Buyer for food processor
Buyer for feed processor
Buyer for industrial materials manufacturer
Country elevator operator and manager
Terminal elevator operator

FIGURE 7-1. "Mountains" of grain come from our western wheat ranches. (Courtesy, International Harvester Co.)

Traffic and transportation specialist
Quality control supervisor for food processor
Quality control supervisor for feed manufacturer
Salespersons (All parts of the grain industry need competent salespersons.)
Grain trader—work on the grain exchanges
Market news reporter
Export service worker
State or federal government inspector and grader
Worker in state or federal government regulatory services
Storage and warehouse manager
Buyer for grain merchandiser
Salesperson for grain merchandiser
Buyer for agricultural industry
Finance and insurance specialist

WHAT QUALIFICATIONS AND ABILITIES ARE NEEDED?

The answer can be very brief and concise. The kinds of positions just listed are exactly the types for which agribusiness education prepares you. Such education allows you to obtain the necessary knowledge in the combination of business and agriculture.

When this is followed by practical on-the-job training and a few years of actual experience, you should be able to carve out a successful career in any one of the positions listed.

In addition to formal education, plus in-service training and on-the-job experience, you should also know farmers, as well as grain and your part of the grain trade business. That combination, plus hard work, should bring success.

So let's examine grain agribusiness in some detail.

HOW WE USE GRAIN

For our convenience, let's make this general classification of our use of grain:

1. *Feed Grains:* These include corn, oats, barley, sorghum grains, and rye.
2. *Food Grains:* These consist primarily of wheat and rice with relatively minor amounts of each of the preceding.
3. *Oil Seeds:* Some contain edible oils; others, non-edible oils. A major portion is used for industrial products. We can include cottonseed, flaxseed, soybeans, castor beans, sesame, safflower.

Keep in mind these principal uses of our vast production of grain crops: food, feed, seed, industrial uses, export.

Here's the way *wheat*, our most important food crop, was used in a recent year:

Domestic Uses

Food .... 900 million bushels
Seed .... 70 million bushels
Industry .... less than 200,000 bushels
Feed .... 55 million bushels

Export
(Will be discussed
  in Chapter 16.) .... 1,000 million bushels

Total Use .... 2,025 million bushels

You will note that wheat is not only a critically important food crop for us here at home but also one of our major exports (in the form of grain or flour). Actually, each of the uses listed gives rise to other agribusiness enterprises.

SPECIALISTS GET GRAIN FROM FARMER TO USER

Grain passes through many channels or agencies that handle, deliver, process, store, and direct its flow. These agencies of the grain industry include:

1. Country elevators and handlers or truckers. These are the primary farm outlets.
2. Brokers, agents, and dealers.
3. Subterminal, terminal, and port elevators which are the secondary farm outlets.
4. Processors of food, feed, and industrial products.
5. Wholesalers and retailers.
6. Exporters.
7. In recent years storages owned or controlled by the Commodity Credit Corporation.

Grain may pass through any number of these channels, depending on its ultimate disposition and other factors.

Now for a closer look at some of these agencies.

Country Elevators: These elevators purchase most of their grain directly from farmers and then sell it later to secondary elevators or to processors. Whenever you drive through the grain-producing regions of our country, you will see many country elevators. They serve relatively small local areas. They perform essential marketing services, such as grading, sorting, storing, and mixing, as well as selling whole grain to larger marketing agencies. Some whole grain is resold to farmers, while some is sold as ground or mixed feed.

Subterminal Elevators: These differ from country elevators in that they receive 50 percent or more of their grain from other elevators or agencies and the remainder from farms. Although they are smaller than terminal elevators, they can store 100,000 bushels

FIGURE 7-2. Kansas City terminal elevator with rail shipping facilities. (Courtesy, Bunge Corporation)

or more. Usually you will find them located in markets that do not have active grain exchanges.

Port Elevators: Located near water transportation, they handle grain that is primarily for export. They may store large amounts, since their storage capacities may equal those of subterminal or even terminal elevators.

Terminal Elevators: These receive nearly all their grain from country and subterminal elevators or from agents, dealers, and brokers. Usually grain comes to the terminals by rail, although some may come by truck or by water transportation.

Terminal elevators grade, price, and store their grain for later sale primarily to processors and exporters. Each year they purchase nearly 50 million tons of grain from farmers, country elevators, brokers, and agents. They sell these millions of tons to food processors, millers, oil seed crushers, exporters, and many other types of users and consumers.

FIGURE 7-3. Terminal elevator with automatic car unloading facilities that can dump a 2,000-bushel railroad car about every 7 to 8 minutes—located at Topeka, Kans. (Courtesy, Farmers Union Cooperative Grain Marketing Association)

Do You Know Farmers?
Do You Know Grain?

If you do, training in agribusiness should qualify you for earning a top "spot" in elevator operation and management. This type of work offers a good career for farm-reared youth who want to work near home. And if successful there, they might, if they wish, move on to even greater responsibilities in the large grain terminals.

PROCESSORS AND MANUFACTURERS

We can group grain processors and manufacturers into three main divisions: food processors (millers), animal feed manufacturers, and manufacturers who make industrial materials from grain. These processors and manufacturers buy their grain from the elevators.

Food: Food from grain comes to us from the flour milling and baking industries. Wheat is the principal grain used, and the products reach us as flour, flour mixes, cereals, bread, and other bakery products.

Significant amounts of corn, oats, barley, and rye are also converted into food products and beverages.

Feed: Processing feed grains into accurately mixed formula feeds for livestock and poultry is another essential agricultural industry. You can read about its many career opportunities in Chapter 8. Feed processing plants are widespread throughout our land. Many are small enterprises serving local areas—another career possibility for those who want to work near home. Other plants are large, serving national or even international markets. Feed manufacturing is a growing business, keeping pace with our increasing national comsumption of meat, poultry, and dairy products.

Industrial Materials: Although a relatively small percentage of our total grain production goes into industrial materials, industrial uses are very important. They are increasing in number; much research is under way to discover new uses. This constitutes a challenging field, and both commercial and government enterprises employ agribusiness specialists.

We Use Grain for Many Purposes in Industry

A detailed list of the many ways in which grains are used in industry would be quite long. But we can look at these few examples.

Each year industry processes 60 million bushels of corn for use in paper sizing, adhesives, building materials, and explosives.

Each year we produce about 5 million pounds of starch, and 95 percent of it comes from the corn wet-milling industry. Cornstarch is used to make paper strong and tough. Paper and paper board take about a billion pounds a year. The textile industry uses some 300 million pounds of cornstarch annually.

Each year some 40 million bushels of grain go into antioxidants, polymers, plasticizers, packaging films, and surface-active agents.

Oil from seeds, such as linseed oil from flax, is used for paints, varnishes, floor coverings, and lubricants. Soybeans and soybean oil have a multitude of industrial uses.

Even the lowly corn cob is converted to industrial use in making 100 million pounds of furfural each year—a raw material for the manufacture of nylon.

THE GRAIN INDUSTRY SUPPLIES MANY BUYERS

You can see that the grain industry supplies many customers. Some purchase grain directly from farmers and ranchers; some buy from country elevators; some buy from subterminal elevators. Exporters may get their supplies from port elevators. The *terminal* markets are the most active meeting places of buyers and sellers. There the *grain exchanges*, and the marketing agencies working with them, buy and sell a large portion of our annual grain production.

GRAIN EXCHANGES

Try to visit a grain exchange. You would find it a thrilling experience. Exchanges are a powerful force in moving the mountains of grain through marketing channels. You would see much action at an exchange—futures traders in the wheat "pit," making

FIGURE 7-4. Trading in the futures pit. (Courtesy, Kansas City Board of Trade)

FIGURE 7-5. Modern exchange room with lighted display. (Courtesy, Chicago Board of Trade)

big deals through motions and gestures. Their hand signals show amounts and prices offered and bargains closed.

You'll see the giant-size board, with its quick-acting attendants posting bids and offers for future delivery so that all may see what is happening to prices. Also these prices are printed in newspapers and broadcast on radio and television. So farmers can know about what price they can expect when they take their wheat, for example, to a nearby country elevator. The price will depend upon grade and quality as well as on the current price at the grain exchange. Farmers are paid in cash when they deliver the grain.

The wheat is shipped by the country elevator operator—usually by rail—to a terminal elevator. But it may not arrive for several days, during which time the price of wheat may drop. For protection, the country manager may "hedge" the transaction and calls the commission person at the exchange and, through this channel, sells a like amount of wheat at the current price for future delivery. By the time the wheat arrives at the terminal elevator, let's say the price has dropped 2 cents a bushel, so there is a loss on that shipment. But now, again through the commission person, the futures contract is repurchased at this lower price. Thus, the loss is offset, and a normal profit is insured, which was computed in the price first paid to the farmer.

Samples of grain are taken when the cars arrive at freight yards near the terminal markets. Experienced graders and inspectors make careful tests to insure that each shipment meets state and federal standards.

You will see samples of the various grains on the cash grain trading tables. Sales made there are for immediate delivery. But no samples are used in futures trading; wheat delivered on future contracts must meet official grading standards as a part of the contracts.

Grain exchanges are located in larger cities of the grain-producing areas. Visitors are welcome. A visit to one of these exchanges, coupled with the literature and pamphlets given out, would heighten your interest in this fundamental agribusiness. The exchange itself never owns a bushel of grain; its primary func-

FIGURE 7-6. Selling grain at an exchange. (Courtesy, Kansas City Board of Trade)

tion is to provide a meeting place and buying-selling facilities for its patrons.

Who Are Members of a Grain Exchange?

A recent annual report of the Minneapolis Grain Exchange answers this for us:

> The members represent all phases of the grain exchange business. Among them are farmers, elevator owners, representatives of farmer cooperatives, millers, maltsters, commission persons, futures traders, food processors, exporters, feed manufacturers and many others.

Market News Reporting May Interest You

This is a profession essential to grain trading. Not only are prices on the exchanges printed and broadcast daily, but also the U.S. Department of Agriculture uses its tremendous mass media facilities. Covering all types of products, these facilities include some 1,500 radio stations, 170 television stations, 1,800 daily newspapers and trade journals, and 200 field reporting offices. Here is a Department of Agriculture statement on the importance of market news about grain:

> Market news made in the downtown building that houses the Chicago Board of Trade is more closely watched in many farm communities across the nation than news from almost any other source.
>
> Prices registered on the trading floor of the "Big Board" set the pace in the marketing of corn, soybeans, and wheat at country points throughout the Grain Belt and beyond.
>
> These commodity prices that farmers are watching are the prices of "futures"—the December, March, May and other "near" and "distant" futures which measure the supply and demand for grain, now and in the months ahead.
>
> It works like this: A farmer gets today's Chicago futures prices—from radio, television, or newspaper. From these prices he can determine the "going" price of cash corn at a nearby elevator, taking into consideration transportation and handling charges. If he thinks the

FIGURE 7-7. Using a helicopter to gather news about crop and livestock management. (Courtesy, University of Florida)

price is favorable, he might decide to sell. If the price has declined, he might hold up. In this way, futures prices serve as base prices in guiding farmers' marketing decisions—whether to sell now, put a crop under price support loan, or hold for later sale.

Regulatory Services at the Grain Exchanges

Regulatory services are performed by a three-agency team: (1) the grain exchange itself, (2) the state government, and (3) the federal government. Exchanges strive to develop close-working relationships with the government representatives. Among the purposes and objectives of the National Grain Trade Council (a voluntary association of grain exchanges and grain trade organizations), you will find these:

1. To promote better understanding among the grain trade, government agencies, and the public generally.
2. To advocate and defend, consistent with public interest, the principles and merits of open and competitive markets for the distribution of agricultural commodities.

FIGURE 7-8. Traders inspecting grain samples. (Courtesy, Minneapolis Grain Exchange)

Each exchange appoints certain of its members to special committees, such as rules, business conduct, price reporting, grain grades, sampling methods, public relations. Each committee performs a type of regulatory service within the exchange.

State Government Inspectors Check Grain Shipments: State government employees, along with their other duties, take samples of grain shipments at the terminal market and detect and report any inferior grain.

Federal Government Representatives Supervise Trading: The federal government, through the Commodity Exchange Authority (CEA) (of the U.S. Department of Agriculture), supervises futures trading on the Chicago Board of Trade, Kansas City Board of Trade, Minneapolis Grain Exchange, New York Cotton Exchange, and 13 other exchanges known as contract markets.

The Commodity Exchange Authority employees enforce limits on speculations in corn, wheat, oats, rye, and soybeans. They strive to obtain accurate registration of prices so as to protect the hedging services of futures markets against unfair or manipulative trading. A supervisor from CEA is usually permanently placed at each major exchange.

The Commodity Exchange Authority watches over 18 agricultural commodities subject to futures trading regulations. During a year there will be a total of some 10 million transactions, with a dollar value of almost $40 billion. And that's only one part of grain trading.

Regulatory services do indeed constitute big business, offering many career opportunities, including a place for you, if you are interested and will prepare for it.

GRAIN STORAGE AND WAREHOUSE SERVICE MAY INTEREST YOU

This is another important part of the grain trade. You might even call it a separate industry. You will be surprised to learn how big it is.

In a recent year we had off-farm commercial storage capacity for nearly 5½ billion bushels. That figure included all elevators, warehouses, terminals, merchant mills, ships under private control, other storage facilities, and oil seed crushers which store grains, flaxseed, or soybeans. It did not include the Commodity Credit Corporation (CCC) storage bins and other facilities under government control.

Of course, many private firms store grains for CCC—some 11,000 in a recent year. Our Department of Agriculture reviews such storage services periodically through sample surveys of 100 or more warehouse firms. The federal government's annual cost for storage—largely for grains—amounts to hundreds of millions of dollars per year.

Storage Facilities Vary in Size

You will see many different sizes of storage facilities. As we said, a country elevator may have space for 100,000 bushels.

Now look at some really big facilities: "Picture a million acres of wheat. That is how much wheat we can store in our elevators." This striking statement comes from American Flours, Inc., of Newton, Kansas.

How many bushels would that be? Figure it out by taking 25 bushels as the yield per acre.

And here's another big grain storage facility—the Garvey grain elevator at Wichita, Kansas, said to be capable of storing over 42 million bushels and to be one of the largest in the world.

Qualified Employees Needed for
Grain Storage Services

The thousands of grain storage enterprises, large and small, widespread throughout our country, need capable personnel for operators, managers, and business executives. They afford another career opportunity for young folks trained in agribusiness.

GRAIN MERCHANDISERS NEED BUSINESS
TRAINED EMPLOYEES

Let's use the word "merchandising" to refer to the services

needed to get the grain from the farmer to the user. We think of merchandisers as individuals, companies, corporations, or cooperatives that actually take title to the grain and have facilities for storing and handling it. Their work includes buying, selling, grading, shipping, and all the accompanying business essentials, such as finance and insurance.

FIGURE 7-9. An inspector at Cargill's Norfolk export elevator operates an experimental television system that accurately and efficiently monitors grain sampling and loading operations. (Courtesy, Kansas City Board of Trade)

Some merchandising concerns are relatively small, such as some of our country elevators. Others are large and serve both domestic and foreign customers.

Take Cargill, Inc., for instance, a grain merchandising corporation of Minneapolis, with an organization that includes over 2,000 grain merchants. Its staff includes *grain scientists, technicians*, and *service people.*

> Cargill men know how to locate best grain source at all times. They know how to move grain from that source to your dock cheaply. They help you buy where grain is in greatest supply, and at the lowest price.

GTA Digest, a publication of the Farmers Union GRAIN TERMINAL ASSOCIATION of St. Paul, Minnesota, gives this interesting description of some of its merchandising operations:

> Single sales in 1,000 bushels are sometimes made by shaving fractions of a cent against a competitor, as by offering better service, by providing just the right quantity and quality discount or a combination of these factors.
>
> Buyers may be at the other end of the nation or across the table in the grain exchange.
>
> GTA handles more country run grain (a grain-originating co-op) than any other organization. It gets its greatest share from the 600 co-op elevators that ship to GTA.
>
> GTA moves large quantities from its elevators across country by truck and train and it is a big user of barges.
>
> [Grain merchandisers] must keep watch of the shifting demands all over the U.S. They must have contacts all over the country, with exporters, and with flour mills and feed processors. They must know where current markets are, where new markets may develop, and get there with the right grain terms and the right place.

Who Buys Grain from the Merchandisers?

Our answer could be a long and detailed list. So we will name only a few of the major users. Note that many of those mentioned are also agricultural industries:

> Flour milling and the bakery industry
> Food processors
> Feed industry (We will consider this in Chapter 8.)

Oil seed crushers and processors
Seed industry
Brewing industry
Distilling industry
Chemical industry
Building materials manufacturers
Paint manufacturers
Manufacturers of various industrial materials
Farmers—livestock feeders, dairy workers, poultry workers
Exporters—one of the biggest purchasers (Our country is the largest exporter of agricultural products, and grain is a principal export item. We will discuss export of grain and other farm products in Chapter 16.)

THINK ABOUT A CAREER IN THE GRAIN INDUSTRY

To move our mountains of grain this industry needs agri-business-trained specialists. It offers attractive opportunities in rural regions, in inland cities, and at seaport locations (in export service).

There is a need for you in the grain industry. The following ad was in a recent edition of *Milling and Baking News*:

WE NEED—IMMEDIATELY

Flour millers   Corporate accounting managers
Second millers   Grain merchandisers
Flour milling engineers   Internal auditors
Elevator managers   Account executives
Feed plant managers   Grain trainees
Chemical engineers   Sales managers

AGRI-Associates, Inc.

Chapter 8

THE FEED INDUSTRY

THE FEED INDUSTRY

This industry processes and sells to farmers and feeders some $15 billion worth of feed each year. Farmers raise many of the raw materials used in manufactured feeds and utilize many of them as they are, without their first being processed. But much of our feed grain leaves the farm to return later as processed formula feeds.

The feed industry conducts many closely related activities which directly increase its production of feed. Related activities include:

> Marketing end products of feed—primarily eggs, broilers, turkeys.
> Processing and packing such end products.
> Operating poultry and livestock feed projects.
> Operating chick hatcheries.
> Contract farming, especially poultry and livestock production.

IT IS A GROWING INDUSTRY

Today we have some 6,500 animal-feed processing plants producing over 105 million tons of feed each year, making it our fourteenth largest manufacturing industry. And it will grow in the years ahead. Our burgeoning world population will require more food, especially the high-protein, diet-enriching foods of animal origin. And here at home our family grows larger and uses more animal products. Each year our total consumption is more than 200 pounds of red meat and poultry, 23 dozen eggs, and 140 quarts of milk for each member of our national family.

Sixteen percent of all farm costs nationally are for formula and concentrated feeds. This is the largest farm expenditure.

The Feed Industry Employs Thousands Today

Even today this manufacturing industry employs some 70,000 workers, two times more than 30 years ago. And to that number we should add thousands more—local feed mill workers, custom grinders and mixers, at least 2,000 wholesale feed dealers, over 20,000 retailers, and several thousand hatcheries that sell feed. All are engaged in a vital agribusiness, which is expanding each year to keep pace with our increasing appetite for meat.

FIGURE 8-1. Modern regional mill. (Courtesy, Agway Inc.)

The Industry Will Need More Tomorrow

Here's a message to you from the American Feed Manufacturers Association:

> Whatever your career interest—scientist, researcher, systems engineer, data processing specialist, salesman, communicator, veterinarian, production engineer, merchandising specialist—wherever your inclinations lead you, there is a place for you in this dynamically expanding feed manufacturing industry.

SOME SPECIFIC OPPORTUNITIES IN THE FEED INDUSTRY

Buyer for Feed Manufacturer: This person purchases grain and other feed ingredients. This is a field service requiring contacts with farmers, grain merchandisers, and makers of feed ingredients.

Manager or Assistant at Feed Manufacturing Plant: General responsibility for all plant operations, procurement of materials, in-plant processes, customer relations, and sales belongs to a plant manager.

Manufacturing Plant Operator: Direct responsibility for plant operations—receiving, mixing, blending, use of additives—is the operator's. In short, the plant operator is charged with the

FIGURE 8-2. Mechanization decreases the farm labor required in feeding the dairy herd. (Courtesy, Agway Inc.)

mechanics of changing raw materials into finished feed products according to prescribed formulas.

Feed Salesperson: All segments of the industry—manufacturers, processors, wholesalers, retailers—need competent salespersons. A good salesperson has a knowledge of feeds and of livestock production problems. He or she knows the language of the industry and can serve customers well. He or she makes contacts and develops good relations with wholesalers, retailers, and farmers.

Feed Wholesaler: The wholesaler purchases from manufacturers or processors and sells and ships to retail stores or other large buyers. The wholesaler may also buy feed for groups of feeders—a wholesale service for groups that cooperate to obtain lower prices by consolidating their purchases.

Employee of a Farmer Cooperative: Feed manufacturing and selling are leading co-op activities. Hence trained people are employed for procurement, plant management, sales promotion, selling, customer relations, and for the many services needed by concerns that handle the product all the way from its source to the final purchaser.

Manager of a Chick Hatchery: There are hatcheries in all parts of the country. Feed manufacturers own and operate many of these hatcheries. This tends to increase sales of their feed and to give them a cushion of an assured market. Managing a hatchery requires contacts with suppliers of fertile eggs; careful in-plant operations; and the development of sales, advertising, shipping, and transportation, along with good customer relations.

Feed Specialist with Chemical or Pharmaceutical Company: Pharmaceutical companies are one of the largest employers of agricultural college graduates. Their chief desire is to obtain agricultural chemists, biochemists, biologists, and other agricultural scientists. But they employ agribusiness-trained specialists also, who can help them contact producers and discover the need for new products.

Manufacturers of farm chemicals employ specialists for contacts with farmers, for demonstrations, for sales promotion programs, for contacts with agricultural colleges and for similar services that help promote the use of feed additives.

Manager of Retail Feed and Supply Store: Being such a manager requires direct contact with feed users, a knowledge of feeds, and familiarity with animal production problems. It also requires good business management and the ability to employ all the essential business services—accounting, cost control, merchandising, store management, advertising, pricing, and aggressive selling.

Opportunities for qualified feed store managers exist in every state.

Employment with Related Industries: The feed industry, as is true with nearly every major industry, is serviced and supplied by many closely related industries. Training in business and agriculture will also qualify you for employment with industries that manufacture items like these:

> Bags
> Feed milling equipment
> Pharmaceuticals and chemicals
> Packaging equipment and materials
> Pelleting machinery
> Feed additives
> Grain cleaning equipment
> Portable feed mixers and grinders

QUALIFICATIONS FOR EMPLOYMENT IN THE FEED INDUSTRY

As you read the preceding list, you will realize that a formal education in the combination of business and agriculture is almost essential. Certainly it is a firm foundation for a career in any part of the feed industry.

In college, these subject areas give basic preparation:

> Courses in animal breeding
> Courses in poultry breeding
> Study of animal nutrition (seems basic to almost all positions in the feed industry)

Several colleges afford opportunity to specialize in the feed manufacturing business. Curricula have been developed in cooperation with the industry. Courses in the following are considered desirable:

Engineering
Business administration
Nutrition
Animal and poultry breeding
Biology
Chemistry
Feed technology
Feed mill construction

Knowledge of the science of animal nutrition is important for all types of work in this industry.

Knowledge of grain, its qualities, standards, and grades, is an essential working tool. But modern formula feeds have many ingredients in addition to grain. Some of these additives are for disease prevention and health promotion; some speed up animal growth; some improve feed utilization. New discoveries in nutrition, new additives, and new methods appear almost every day. Your success may depend on your keeping in step with this rapid progress. Of course, you will need on-the-job training—some months, or years, of experience together with the benefits of association with successful administrators and executives.

Today's modern feed industry produces scientifically blended feeds, based on careful research.

YOU MAY FIND YOUR CAREER IN THE FEED INDUSTRY

Certainly the industry offers many opportunities. It's a growing business. It quickly adopts advances in technology to produce today's modern formula feeds.

What Are Formula Feeds?

There are many variations—perhaps some 1,500 different combinations of ingredients. In recent years, research in animal nutrition has made rapid progress; its discoveries have required the feed industry to develop new, scientifically blended products.

To prepare yourself for a career in the feed industry, you will need a good background in the principles of animal nutrition. Whether you become a salesperson, sales manager, or business manager, you will need a good foundation in this basic science.

Formula feeds have many ingredients, mixed and balanced to promote health and productivity in livestock and poultry. Chemists, animal breeders, poultry specialists, biologists, and other scientists in public and private agencies have cooperated in developing the knowledge essential to the feed manufacturing industry.

Probably over half of the ingredients used by feed processors come from these agricultural products and their by-products: corn, soybean cake and meal, oats, millfeed and screenings, meat scrap and tankage, wheat, alfalfa meal, and cottonseed cake and meal. To these major ingredients, processors add a wide variety of vitamins, minerals, drugs, and antibiotics.

Our program in animal nutrition and the development of high efficiency feeds have given us remarkable gains in our animal agriculture.

Feed Manufacturing Operations

First Comes Procurement or Purchasing: This means contacts with farmers, elevators, warehouses, or other sources of supply. It takes persons who know grain—its qualities and grades, test results, and standards, as well as appropriate prices. A large concern, with several processing plants, may have a central purchasing department. This gives the advantage of larger quantity purchases and usually means lower costs.

Purchasing or procurement is an attractive field for agribusiness graduates who have farm backgrounds.

Then Come Plant Operations: Grain and feed ingredients are received in bulk at the mill. They are unloaded and transferred by air-pressure units—or similar automatic facilities—to storage bins and work bins. In fact, such transfer facilities are used for almost all the handling procedures.

Here's a good description of the operations of a relatively small co-op feed mill:

> This new Fredericton (New Brunswick, Canada) branch of the Maritime Cooperative Services, Limited, presents a new concept in feed mixing in that it is a pre-weigh, pre-grind combination, semi-automatic mill. This means that the grain portion of a ton of mix of feed is drawn from the storage bins, weighed, and ground. This insures the freshest possible feed. This mill has bulk storage of 17,500 bushels with an eight hour capacity of approximately 50 tons of mixed feeds or concentrates. It is equipped with a one ton mixer and surge bin, a pellet mill and cooler, molasses mixer, hammer mill and crusher, as well as the latest in handling equipment for unloading grain and transferring ingredients and feed in the mill.

Perhaps you would classify this particular mill as a rural enterprise, serving a local area. However, it is but one of the members of a co-op organization that affords the Maritime Provinces of Canada the most modern feed processing facilities.

Feed manufacturing enterprises vary greatly in size; some are smaller than the Fredericton Mill just described.

Today, modern feed milling is done largely on a regional basis with many of the older and larger feed mills out of existence. The regional mills are located near the users in order to reduce transportation costs on the finished feed. Many, if not most, of these are equipped with semi-automatic batching systems. Some are controlled by punch cards and some by small computers.

Additives and Supplements Are Used: Today's high efficiency feeds contain a variety of supplements and additives—agricultural by-products, meals, oil cakes, vitamins, chemicals, medicines, and drugs. Accordingly, related industries furnish such materials for our almost unlimited variety of animal and poultry feeds.

By-products used include brewers' yeast, sugar beet tops, dried beet pulp, dried citrus pulp, and similar items. After oil seeds are processed and their oil is extracted, a residual feedstuff containing 35 to 50 percent protein is processed into cake and meal. Sources of such valuable feed supplements include soybeans, cottonseed, peanuts, and safflower.

Trace minerals—cobalt, manganese, zinc, boron, copper, and iodine—are important additives also.

Medicines and Drugs Are Employed: Within the last 20 years or so, the feed industry has introduced preventative medications into formula feeds, as well as adding many growth- and production-promoting preparations. Concerns making such products operate well equipped laboratories, manned by highly trained nutritionists and research scientists.

You will be surprised to learn that pharmaceutical and chemical companies are major employers of agricultural graduates. That is because new additives are becoming so valuable in animal and poultry production.

Affiliated Industries Increase Employment Opportunities

All of these new industries contributing to feed manufacturing mean more career opportunities for agribusiness specialists. Makers of these new additives need not only chemists, biochemists, and research scientists but also business-trained, well qualified persons to meet farmers, feeders, and feed manufacturers. They need people who know feeds and feeding and animal nutrition, people who can know and understand the goals and objectives of the producers of meat and poultry products.

FEED MANUFACTURING IS A HIGHLY DIVERSIFIED BUSINESS

Some manufacturers make and sell feed only. But many engage in related activities such as processing and selling the end products their feeds have produced, primarily to increase sale of their feeds.

Feedstuffs,[1] the weekly newspaper of the feed industry, quite recently reported on such "extra-curricular" activities. This report

---

[1]*Feedstuffs,* The Miller Publishing Co., 2501 Wayzata Blvd., Minneapolis, Minn. 55405.

shows the value of each such activity in relation to the total business of feed manufacturers who carry on one or more of them:

Activity	Percent of Total Business
Marketing end products of feed	33.9
Processing of end products (primarily eggs, broilers, turkeys) ..	22.8
Contract production (eggs, turkeys, hogs, beef cattle)	46.9
Direct operation of feeding establishments	37.8
Operation of hatcheries	22.8

FIGURE 8-3. An "end product" of the broiler industry—an outdoor barbecue. (Courtesy, Maine Department of Agriculture)

FEED INDUSTRY DEVELOPS CONTRACT FARMING

Feed processors in our southern states were the first to begin furnishing feed to broiler producers under some type of contract. This integrated method proved quite successful and gave great impetus and rapid growth to broiler production.

The broiler enterprise grew rapidly between 1955 and 1961, especially in Alabama, Arkansas, Georgia, Mississippi, and North Carolina. And throughout our nation there has been a tremendous increase in broiler production. Almost 2½ billion broilers were produced in the United States in a recent year, more than three times as many as in 1950.

See How Contract Farming Has Developed in Arkansas

About 95 percent of Arkansas' poultry producers operate under contract with feed processors or feed suppliers. Some growers furnish labor, litter, housing, and equipment; contractors supply the feed and, in some cases, supply the birds. Growers may be paid on a per pound, per head, or per dozen basis, or the price may be based on the gain in weight during the feeding period. Other growers may be hired on a weekly labor basis. In that case the contractors' birds are kept on farms which they own or lease.

Soon after contract farming became general, the feed supplying contractors began to process and sell the poultry products, thus further integrating their businesses.

Under various forms of contracts, poultry producers and many livestock producers have, in a sense, become subcontractors—and sometimes partners—of the feed industry.

You Will Find Contract Farming in Livestock Production Also

A study by The Iowa State University a few years ago showed that some 18.6 percent of the total sales of livestock feeds in Iowa,

Illinois, Missouri, Minnesota, Nebraska, and South Dakota were made under some form of financing or contracting program.

These programs ranged from loose financing arrangements, with no supervision of the farmer's production operations, to highly integrated programs for livestock supply and final marketing as well as for the feed and other production supplies. The one common characteristic of the programs is that they provide a farmer with credit for the feed to be used over a specified time period (or livestock production cycle) in return for which he agrees to use the manufacturer's (or dealer's) feed during the period of the contract.

THE FEED INDUSTRY NEEDS AGRIBUSINESS SPECIALISTS

The Iowa study also revealed the need for better managerial direction and cost control—careful examination of charges, prices,

FIGURE 8-4. Many feed manufacturers hatch and raise chickens, thus assuring a market for their feed. (Courtesy, Agway Inc.)

FIGURE 8-5. Operator at electronic control panel. Activated by data cards containing feed formulas, this equipment automatically selects and regulates deliveries of ingredients to feed blenders in predetermined amounts and in the prescribed sequences. (Courtesy, *Feeds Illustrated*)

and the cost of specific operations. This industry has grown so fast—some 300 percent within a 20-year period—that many firms are still understaffed.

The feed business is diversified. Companies carry on many related activities. They must maintain close contacts and good relations with farmer-producers. Employees, who know animal nutrition, biology, feeds, and feeding, are essential to the success of the enterprises. Feed manufacturers, dealers, and contractors operate on small margins. They need efficient cost control and business management.

Linear Programming Now Employed

Formulation, inventory control, accounting, invoicing, and payroll are all functions that are now handled by computer.

Formulation is still linear programmed for least cost either on a single feed basis or on a multiple feed basis (to make best use of ingredients on hand when the inventory mix is not well balanced).

Take advantage of your opportunity to study linear programming methods while you are in college.

SUMMARY

Conditions in the feed industry—rapid expansion, increasing diversification, inadequate business practices of many firms—clearly reveal the need for agribusiness graduates and offer them promising careers.

The need for management personnel is so urgent that feed industry people now serve actively on advisory committees of our agricultural colleges. They help shape curricula to combine business practices with animal nutrition sciences and technology. No doubt you will find such a study combination at your own state college, because the feed industry is active and expanding in almost all our states.

You would be welcomed by this industry.

Chapter 9

THE MEAT AND
LIVESTOCK INDUSTRY

THE MEAT AND
LIVESTOCK INDUSTRY

> In no field of industry has a greater miracle been
> wrought than in that segment which produces the meat
> for the American table.[1]

A miracle indeed! The American meat and livestock industry
provides about one-third of all the meat produced in the whole
world. Efficient, well organized cooperation of producers, feeders,
marketers, manufacturers, distributors, and retailers puts 35 billion
pounds of meat on our tables each year—160 pounds of red meat
for each of us.

THOUSANDS OF PEOPLE TAKE PART
IN THIS BIG JOB

There are about 315,000 people working in this industry.

Think of the many activities needed to get this huge amount
of meat to us:

1. Purchase from farmers and feeders of some 130 million cattle, calves, hogs, sheep, and lambs.
2. Purchase of chicken, turkeys, geese, and ducks.
3. Transporting of live animals and birds to stockyards, terminal markets, and local markets.
4. Sales to slaughterers and packers.
5. Slaughter and primary processing.
6. Sales to processors and manufacturers.

---

[1]Bertrand B. Fowler, *Men, Meat, and Miracles*, Julian Messner, Inc., New
York, N.Y.

7. Manufacture and processing of finished meat products and production of by-products.
8. Distribution and transportation to retail stores.
9. Retail sales.

A HIGHLY DIVERSIFIED INDUSTRY

The meat and livestock industry is highly diversified. It produces many important by-products. It is active in all parts of our country. Some of its operations are going on right near you.

Look back from your dinner table at the whole wide scene of this vital industry. You'll see retail stores and meat markets, wholesalers, distributors, processors, meat product manufacturers, stockyards and terminal livestock markets, professional buyers, sellers and dealers, transporting and distributing agencies, feedlots, farms and ranches, plus all the protective services of inspection, grading, and labeling.

You will find many opportunities in the meat and livestock industry.

One large meat processing concern with many employees around the world has this to say about its company:

> At Wilson Foods, people are important. The corporation employs some 12,000 persons worldwide. They range from livestock buyers to plant production line workers; from research food scientists to economists and computer technologists; from salespersons and merchandisers to clerks, spice blenders, and secretaries; from attorneys and engineers to analysts, skilled butchers, accountants, writers, and brand managers.
>
> As a college graduate you will find a wide range of opportunities within the company—positions that will offer you challenging, interesting assignments, personal fulfillment, and a promising future.

IT'S A BIG AGRIBUSINESS INDUSTRY

The U.S. Bureau of the Census and the U.S. Department of Agriculture give us detailed information of the size, scope, and achievements of the meat and livestock industry.

In a recent year, over 30 million head of cattle were converted into meat; more than 80 million hogs were processed to give us 15 million pounds of pork products; nearly 200 million pounds of lamb and mutton were prepared for us as well as 55 pounds of poultry meat for each one of us.

Many Agencies and Services Participate

Here are some of the many agencies and services that took part in this big job: 5,000 meat product manufacturing establishments, 5,000 wholesalers of meat and meat products, and 2,000,000 retail meat markets.

Transportation Services: Today's meat industry depends on rapid, well engineered transportation of a highly perishable product. You will appreciate the importance of efficient transportation when you realize that some two-thirds of our livestock is produced west of the Mississippi River, while about two-thirds of the consumers live east of the Mississippi.

CAREERS IN THE MEAT AND LIVESTOCK INDUSTRY

We will list a number of positions that require training in business and agriculture and, in the following pages, give you a description of the type of work performed in such positions.

But first, be assured that there are many promising career opportunities in this most important agribusiness. Consider this message from one of our major meat packers:

> There are hundreds of different types of jobs in the more than 300 Swift and Company plants, offices and other units throughout the nation.
> Swift needs a constant flow of persons with a wide variety of skills, training, and background to maintain its working force and to permit growth of the business.

Specific Positions in the Meat and Livestock Industry

Buyer of Livestock and Poultry: Persons capable of rendering this service are required by many parts of the industry. In such a position you might buy from farmers and feeders, for slaughterers and packing plants, for manufacturers and processors, for wholesale distributors and retail stores.

Manager or Assistant at a Feedlot: Feeding cattle for the market is carried on in every state. Qualified employees soon advance to responsible positions.

Salesperson: All parts of the industry need salespersons. Sales are made to many types of clients, from producers to consumers.

Manager or Assistant at Processing or Manufacturing Plant: This is an in-plant type of position offering a variety of technical and management opportunities.

Transportation Specialist: Live animals and birds are subject to injury in shipment, which causes serious losses. Meat products are perishable and need expert transportation methods. Transportation is a special field with opportunity for valuable service to employers.

Service Worker with Wholesaler: This position requires selected personnel for purchase and procurement; for warehousing and storage; for selling, sales promotion, advertising, and the development of customer good will through service to retail dealers.

Cattle Buyer at Livestock Market (Stockyards): Success requires good judgment, a thorough knowledge of cattle, and an alert business sense, as competition among buyers is keen. You will become skilled in your job in this rapid-action special service through experience. You may have many different types of clients because all classes of cattle purchasers use the livestock market.

Cattle Seller at Livestock Market: You may sell for these clients as well as buy for them. In selling, too, you must match wits with sharp buyers to make your bargains.

Commission Persons: These persons may be the selling agents for farmers or feedlot operators, and sometimes they may buy feeder stock for such operators. They are paid a commission for their services.

FIGURE 9-1. Feeder steers sold at auction sponsored by Fall River Big Valley Cattle-men Association at McArthur, Calif. (Courtesy, USDA)

Inspector or Other Regulatory Service Workers: Such a position may bring employment with federal, state, or city government. Regulations, standards, grades, and labels are rigidly maintained by qualified specialists.

Employee of Large Meat Packing Corporation: Among the positions with a large meat packer which the college graduate may work toward are:

> Production manager
> Marketer
> Sales administrator
> Data processor
> Cattle buyer
> Labor relations representative
> Personnel director
> Assistant to personnel director

Manager or Assistant at Retail Store: This position involves all the techniques and modern methods of retailing, the end of the chain from producer to consumer.

Worker in Business Services: The meat and livestock industry needs many special services for which agribusiness-trained people are preferred. These include:

> Marketing
> Cost control
> Accounting
> Sales promotion
> Evaluation of new products
> Pricing
> Advertising
> Consumer relations

With its great variety of requirements for well trained professional workers, the Meat and Livestock Industry has a place for you if and when you have prepared for it.

Qualifications for Careers in the Meat and Livestock Industry

College Training: Preferred positions are offered to those with college training, especially to graduates. Formal education and specialization in agricultural sciences and business administration will prepare you for a career in agribusiness or industry.

In agriculture, majors should be chosen in animal sciences and animal and poultry breeding. Courses in animal judging and poultry judging will be of great value. Take as many electives as possible that relate to livestock feeding, meat processing, livestock and poultry marketing, and packing plant operations.

In business administration, try to include courses in accounting, cost accounting, marketing, commercial correspondence, economics, statistics, sales management, sales promotion, and advertising.

On-the-Job Training: This is almost always necessary. A college curriculum cannot possibly cover all the essential practices or customs of a given occupation—certainly not for the many specialized activities of this industry.

Almost all employers, large or small, provide some kind of in-service guidance, because that increases your value to them.

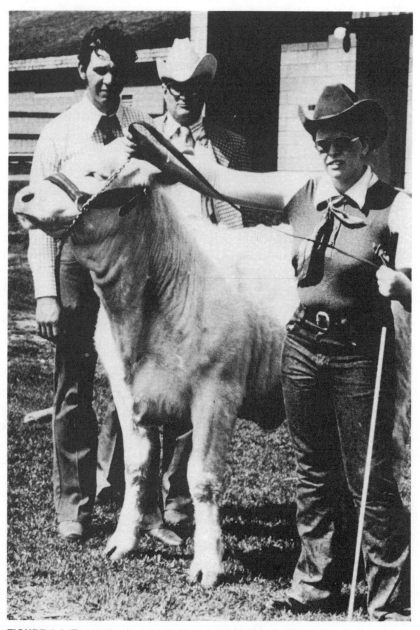

FIGURE 9-2. To prepare for a career in the meat industry you should take a course in judging beef animals. (Courtesy, S. Pendrak, SUNY–Cobleskill)

Experience: Whether weeks, months, or longer periods are necessary depends on the particular position. But this stage of your training may be most valuable. It is "first-hand" and "down to earth" and prepares you for the road to advancement.

Throughout the rest of this chapter we will point out other desirable qualifications as we discuss the various components of the industry and their operations, functions, and customs.

WHERE DO OUR MEAT ANIMALS COME FROM?

Beef Cattle

The meat on your table has made quite a journey to get to you.

Take beef cattle as an example: Although beef cattle are raised in every state, a major proportion start their lives on the ranches of the West and Southwest. There they live primarily on grass and hay. Over half of our farm land is in grass—some 500 million acres plus 300 million acres of uncultivated range land. The grass from this great area can be marketed only through live-stock.

After a summer or two on grass, the cattle are sold and shipped to "feeders," who fatten and "finish" the animals on grain and other concentrated feeds. These highly nutritious, scientifically blended feeds rapidly put weight on the cattle and make our beef more tender and flavorous than "grass-fed" beef.

Cattle finishing in the feedlots is a relatively new and highly specialized operation in our livestock industry. It is carried on now in all sections of our country. About one-half of our states are now considered major cattle feeder states.

The number of our feedlots increases each year. At the same time they get larger and more highly capitalized, mechanized, and "automated." Their annual capacity is greater. They feed thousands of animals each year; some feedlots handle 20,000 and some up to 100,000. In most cases, feedlot owners buy, feed, and sell their own stock. Some afford a custom service—feeding cattle for other owners. Some feedlots are owned and operated by feed manufacturers to provide an assured outlet for their feed products.

Our feedlot operators were feeding nearly 12 million head of cattle at the beginning of a recent year.

Hogs May Take a Shorter Route

Hogs reach the packing house more directly than cattle. Usually hogs are born, fed, and finished for market on the same farm. But some go to the feedlots, and we now have some large-scale hog-feeding enterprises.

Although hogs eat some grass, they must have grain or other concentrated feeds for proper growth. Accordingly, the feed-grain areas also produce the greatest numbers of hogs, although some are raised in all farming areas. Hogs mature quickly and may be ready for market in six to eight months after farrowing. Producers can expand or decrease their output easily to bring production more closely in line with demand. Cattle production takes much longer. Hence, the beef supply cannot be changed so quickly as the pork supply.

Manufacturing and processing of pork and pork products constitute a big segment of the entire meat industry. To satisfy our appetites, processors convert nearly 80 million hogs into such products annually.

Sheep and Lambs from the West and Southwest

Most sheep and lambs come to us from ranches in the West and Southwest, but some come from farms in the Midwest, South, and East. Lambs are good foragers and grow quickly if they have ample feed. Many are ready for market after one summer of grazing. Some that may be too light for slaughter are sold to feeders, who fatten the lambs for three or four months. The meat industry processes about 12 million sheep and lambs each year.

WHO BUYS LIVESTOCK FROM OUR FARMS AND FEEDLOTS?

Livestock raisers and feedlot operators have many customers—a wide choice of buyers for their finished animals.

A study made in the 14 Corn Belt states revealed that producers in those states sell to, or through, 12,000 dealers and truckers, 2,900 local meat dealers who do some slaughtering, 1,000 auction markets, 1,000 cooperative associations, 600 direct buying packing plants, 300 concentration yards, as well as 550 commission firms operating on 26 large public markets.[2]

Try to Visit the Stockyards

A large proportion of our livestock goes to the stockyards at terminal markets. There you can see how quickly sales are made, how efficiently animals are handled, how buyers and sellers make their contacts, and how prices are established. In brief, you would learn a lot about large-scale livestock marketing.

Perhaps there is a stockyard near you. We have hundreds of local public livestock markets and a dozen or more major ones. The large terminal markets (stockyards) are located near the livestock-producing areas—Omaha, South St. Paul, Sioux City, Kansas City, East St. Louis, St. Joseph, Denver, Fort Worth, Indianapolis, Wichita, and Oklahoma City. These locations assure a year-round supply of livestock for the market.

At the stockyards you'll see a sequence of operations like this:

1. Railroads and trucks haul animals to the unloading chutes.
2. Stockyard workers receive the animals, then feed and water them; veterinarians inspect the animals.
3. Stockyard crews or sales agency persons drive animals to the sales pens, where the animals are allowed to rest. The sales agency feeds, sorts, and classifies the animals and prepares them for sale.
4. Sales agency persons and buyers negotiate the sales and then the agency persons drive the animals to the scales.
5. Stockyard representatives weigh the animals in the presence of both buyer and seller. This is done on scales periodically checked and tested by the U.S. Department of Agriculture.

---

[2]As reported by Armour and Co., Chicago, Ill.

FIGURE 9-3. Large livestock terminal market and stockyards. (Courtesy, USDA)

6. Title to the livestock passes from seller to buyer when the animals are weighed.
7. Stockyard personnel take the animals to the holding pens and care for them until the buyer requests delivery.
8. The stockyard ships or delivers the animals to the buyer.

The sales agency pays the livestock producer after deducting its commission and the charges for the stockyard services.

Terminal markets and their yards are essential to the meat and livestock industry. They are concentration centers, where buyers can get the particular grade and weight of animals they need. These big markets and also the smaller public markets do the following:

1. Help move animals from producing areas to the consumer markets.
2. Bring buyers and sellers together in active competition.
3. Stabilize and establish prices.
4. Maintain safe, orderly, and efficient operations and are subject to close inspection and supervision by federal and state governments.

5. Reflect consumer preferences for meat and thereby contribute to the improvement of livestock.

You'll See Fast-Action Buying at the Stockyards: There are many types of buyers and much keen competition. Each buyer tries to get animals for the special needs of his or her firm or client, which may be a local or regional meat packing house, a national packer, a wholesaler, a locker plant operator, a farmer (buying cattle for feeding or stocking), or a livestock dealer.

COMMISSION PERSONS ARE IMPORTANT

Consider the commission persons at a large terminal market. They are the selling agents for the producers or feedlot operators. When the animals are unloaded from trucks and trains at the stockyards, the commission person invites buyers into the pens to make their bids or offers. When he or she and a buyer reach an agreement, the sale is completed. Then the animals are weighed, and cash payment is made on the basis of weight and the agreed-upon price. The producer or feedlot operator pays the commission person a fee for his or her services.

So you can see that a professional commission person is more than a salesperson in the ordinary sense. He or she must know the product and the buyer and must establish the value of the product. Success depends, in large measure, on the ability to get the best the day's market affords. Experience, skill, and market knowledge of the commission person afford a most valuable professional service to producers and feeders.

The commission person may perform many other services:

> The livestock commission persons serving on the public markets are more than salespersons. They are part of a team working with livestock producers the year round. Very often they're in on the purchase of feeder stock, work with farmers and feeders through the growing and feeding process, keep abreast of trends in the trade, and counsel in the marketing decisions.[3]

---

[3] Swift and Company.

CONSIDER THE MEAT PACKERS

Many of us think all the meat packers are huge corporations. Of course, some are very large and serve our whole country. But many are relatively small. In our west coast cities—Los Angeles, for example—you might find several dozen packers in addition to the major ones. But whether the packers are large or small, activities such as the following are essential: procurement, manufacturing and processing, selling.

Procurement: This means purchase of livestock from markets, stockyards, dealers, feedlots, farmers, or contract producers. Purchasing requires persons who know livestock and whose business training, coupled with experience, gives them the keen sense of values and the bargaining skill needed for their exacting function. Agribusiness graduates who were raised on farms may have an advantage in livestock purchasing.

Manufacturing and Processing: These terms cover the many plant operations needed to convert live animals into meat, meat products, and a great variety of by-products and related products. Following are some principal plant operations:

Slaughtering is done by humane methods. Carcasses are cleaned, split in half, and then trimmed, washed, covered with muslin, and moved into the chill room. Some are shipped by rail or truck to distant customers or to the packer's branch houses; others are processed further within the plant.

Processing includes many operations that convert raw materials into dressed meat and finished meat products. And numerous valuable by-products are developed, as well as almost countless related products—lard, fats, leather, glue, soap, glycerine, plant foods, animal feeds, industrial materials, drugs, pharmaceuticals, and many other products. All parts of the animals are used—bones, horns, hoofs, glands, viscera, blood, and sinews.

Today with our modern methods of temperature control and product protection, much meat comes to us from the packer through the retail store as fresh meat. But large quantitites are cured, canned, frozen, or made into sausage. About 40 percent of

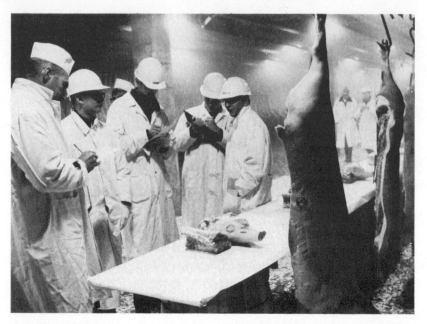

FIGURE 9-4. Your class in meat grading may visit a local cooler. (Courtesy, Iowa State University)

our pork is cured as ham, bacon, or picnics. Beef also is cured and some is dried; both corned beef and dried beef are quite popular.

We're also using more meat-in-a-can products—such time savers as beef stew, chili, corned beef hash, ham, roast beef, luncheon meat, and a host of other meat products.

Sausage of many kinds is perhaps the most popular processed meat; frankfurters—known by many names—comprise nearly one-third of all the sausage meat sold in the United States.

Selling for a Meat Packer: The diverse products of a national packing house are sold to many types of customers—to other processors and manufacturers, to wholesalers, retailers, restaurants, hotels, public institutions, and other large buyers.

Business-trained salespersons are needed at the central plant, and even more are required at the branch houses. These branches are located at strategic places throughout our whole nation. One large packer has nearly 300 branch houses. So you may find a sales career opportunity right near your home. Branches are primarily distributing and selling agencies, although some may also do processing.

Prospective salespersons are given special training to learn the company's methods and its various products. Later they may be assigned to work with certain types of customers or to concentrate on selling particular products or by-products. The well diversified business of a major packer can afford you many chances for "specialization" in selling.

STORAGE AND WAREHOUSING

Refrigerated warehouses, owned by the packer or the retailer, form a link between the processor and the consumer.

Warehouses can hold meat safely so that it is always ready for delivery upon call from the retail store. Today, many of our retailers receive meat directly from the packer.

PRICING AND COST CONTROL

The more you learn about the meat industry, the more you will appreciate the need for good business management. Packing plants operate on small margins, and they handle perishable products. Sometimes supplies are uncertain; supplies depend on when the livestock producers have animals ready for sale and when they want to market them.

No wonder meat packers welcome agribusiness graduates who, after in-service training and experience, can qualify for management positions in this basic industry. They need young people, with college training, who can carry on in the years ahead and help the industry continue to solve its problems.

> The packing industry meets the problems of distribution so quickly that meat is never allowed to spoil because there is no market for it. On the contrary, all the meat which is produced moves into consumption at approximately the same rate as animals come to the market.
> No other industry does so much for so little return as the packing industry. For all buying, slaughtering, proc-

essing and distributing operations, the industry's net
profit since 1925 has averaged less than ⅓ of a cent per
pound of meat, or ¼ of a cent per pound of livestock.[4]

MARKETING MAY HOLD OPPORTUNITIES FOR YOU

You may find your career in marketing. It's a broad field and a
vital part of agribusiness—especially in the meat and livestock industry.

What Is Marketing?

It includes "all the processes that occur after the product
leaves the farm—the buying and selling, assembling, transporting,
warehousing, processing, grading, packaging, merchandising—in
short, all that is required to turn a steer into a steak, to make the
wool from a sheep into a gray flannel suit, and to fill the shelves of
a modern supermarket with 10,000 different items."[5]

Buying and selling is a basic marketing activity. A meat product, as it passes from producer to consumer, may be bought and
sold many times. Large firms have special purchasing departments
and sales departments. And there is a variety of marketing positions in their branch houses, as well as in their central plants.

It takes many well trained specialists, managers, and executives to handle the tremendous volume of livestock marketing
throughout our nation.

That's why livestock marketing enterprises are seeking agribusiness graduates. Those who combine the study of livestock
with professional business training can find a profitable and rewarding career in livestock marketing.

INSPECTION AND REGULATORY SERVICES

You will find many employment opportunities in these services for the livestock and meat industry.

---

[4]Armour and Co., Chicago, Ill.
[5]U.S. Department of Agriculture.

FIGURE 9-5. Meat being cut and prepared in a retail meat store. (Courtesy, Swift and Company)

Our federal, state, and municipal governments are all concerned with food and meat inspection regulations, standards, grades, and labeling.

The Meat Inspection Division of the U.S. Department of Agriculture conducts the inspection service for our federal government. Federal inspectors try to make certain that our meat is wholesome. They require the diversion or destruction of diseased animals and prevent the use of impure dressed meat.

Federal inspectors and other federal officials have important duties in the livestock markets. They administer the Packers and Stockyards Act. This law helps to assure free, open, and fair competition and fair business practices in the marketing of livestock, meat, and poultry all the way from the farm to the retail store. One objective is to help producers get true market value for their livestock and poultry; another is to protect consumers against unfair business practices in the marketing of meats and poultry.

You would gain valuable experience by serving as one of the

more than 5,000 federal inspectors on duty in meat and poultry slaughtering and processing plants. These inspectors are busy workers. During a year they certify some 25 billion pounds of meat and poultry as meeting the established standards for wholesomeness and safety.

With cattle, hogs, and sheep, inspection starts with live animals in the holding pens. Each animal is examined; suspected ones are condemned or diverted for further attention.

After slaughter, the inspectors examine each carcass and continue their inspection through each stage of curing, canning, sausage-making, or other processing.

State and municipal governments also perform inspection and regulatory services. Usually the state service is under the State Department of Agriculture. In many states the processors and manufacturers have requested these services. Among other advantages, having their meat items pass inspection enables them to advertise their products as meeting the rigid state requirements. This helps them establish consumer preference for their special brands.

State inspection may include a careful check of the processors' plant facilities, equipment, water supply, raw material sources, and important supplies.

Municipal inspection includes surveys and examinations of supply warehouses, retail stores, delivery trucks, refrigeration systems, locker storage, restaurants, hotel dining rooms, and other public eating places.

If you could see our meat and poultry inspection, grading, and regulatory system in all its stages from the farm to your table, you would agree with this statement by the U.S. Department of Agriculture: "Our meat and poultry inspection system is the envy of the whole world." And it takes trained workers and efficient specialists to make it deserving of this high praise.

RETAIL STORES

There are over 2 million retail meat markets in the United States. Some are small, and some measure their annual volume in millions of dollars. The trend is toward big business—more merg-

FIGURE 9-6. Retail meat store. Good displays and attractive refrigerated compartments result from careful planning. (Courtesy, National Association of Retail Grocers)

ers and consolidations. This trend means increased demand for agribusiness graduates who have the capacity for growth and the willingness to gain experience—to know the business—so they can become managers and administrators.

Retailers are closer to the consumer than either the packers or the producers. The retailers learn what you and your neighbors want and relay this information to their sources of supply. Actually, they write the specifications for the kinds of meat products their suppliers must furnish.

To have what you want is the retailers' problem. To get what you want, large retailers procure products from many sources—from nearby or distant packers, from their own direct purchases at terminal markets and auction sales, and perhaps from their own feedlots. Retailers may procure sausage and processed meats from different sources than the ones from which they buy fresh meats.

Observe the operation of a good retailer. You will see that he or she knows what you want and has it on hand. It is well displayed at the proper place and at the right time and is available in the unit quantities that are convenient for you.

Retail Management Brings Real Challenges

Retail management must be forward-looking. There will be many new developments in the meat industry—new products, new methods of preserving and packaging, perhaps the widespread use of irradiation as a preserving method. Specification buying will increase in response to customer preference. Marketing costs mount up.

Marketing Costs: Together, all stages of marketing now take about 69 cents of our food dollar. Now we spend about 30 percent of our food dollar for meat and meat products. It probably costs the retailers 20 cents a pound or more (wholesale weight) to market their meat. Retail costs include wages and salaries of employees; procurement of packaging materials; product preparation; overhead costs; taxes; return to stockholders on investment; purchase of meat, meat products, and supplies.

If you use all the tools of good management—careful buying, fast transportation and delivery, efficient storage and warehousing,

FIGURE 9-7. Research kitchen for checking the cooking characteristics of different meat cuts and grades. (Courtesy, Swift and Company)

refrigeration, artful and tasteful display—and keep costs down, you can meet the keen competition and win the loyalty of your customers.

It's not an easy job! But it can be a profitable and rewarding one. It's a job that requires fundamental business training and product knowledge. And meat retailing is an important community service. "Americans consider meat the most important item in most meals," reports the U.S. Department of Agriculture.

You Would Serve Your Whole Community as a Retail Executive

Good retail management means much to the whole community. It means wholesome, high quality food products, convenient self-service, unit size quantities, transparent packaging, easy inspection. But even more important, efficient retailing can give us real value for our money and many conveniences and other advantages.

FIGURE 9-8. Small retail meat market offers an opportunity for community service and satisfactory income. (Photo by H. E. Gulvin)

CHALLENGES AND OPPORTUNITIES

This highly diversified industry, which is local, national, and international in scope and which performs many functions and employs thousands, offers opportunities and challenges. Indeed, it is diversified and broad in scope when a major company, like Armour, produces 2,000 food and chemical products.

Armour and Co. sends this message to college students:

> Diversification is the challenge at Armour that now faces the college man or woman in search of a career; a company in transition to pace a world in transition.
>
> The broad scope of activity at Armour enlists the full range of management arts, sciences, and techniques. The company's new approaches to mechanization, marketing, distribution, and communication are opening up careers for college people with a creative view to the future.
>
> A new employee at Armour receives no guarantee of automatic promotion. He or she does get assurance of the chance to explore, develop, and demonstrate his or her potential in one of the most favorable environments for growth in today's business world.

If you could "write your own ticket" for a career opportunity, could you improve on that statement by Armour?

The meat and livestock industry, in all its varied phases from producer to consumer, has many good career prospects for agribusiness graduates who have vision, imagination, and perseverance.

Chapter 10

THE COTTON INDUSTRY

THE COTTON INDUSTRY

Not only do our farms feed us, but also their products clothe us, furnish and decorate our homes, and contribute to our comfort and health.

In this chapter you will learn how cotton, a farm-grown fiber, passes through the channels of agribusiness, where it is made into things we need and use every day, as well as into countless industrial products.

COTTON BOLL →GIN→MILL→ CLOTH, CLOTHING →CONSUMER

THE COTTON INDUSTRY IS ESSENTIAL IN OUR ECONOMY

The cotton industry takes cotton from farmers and, by many activities and services, transfers it to users in our country and throughout the world. It employs thousands of workers and affords many opportunities for college-trained business specialists. *Consider the cotton or textile industry in your career.*

There are over 1 million persons employed in textile mills, with the same number employed in the apparel industry.

COTTON—A GREAT CASH CROP— A MOST IMPORTANT FIBER

Cotton is one of our great cash crops and a most important fiber; it makes up about one-fourth of our total fiber use.

Wool is important also and is preferred for many purposes, as you know. We use nearly 130 million pounds a year—about ½

FIGURE 10-1. Loading cotton for export. (Courtesy, The National Cotton Council of America)

pound for each one of us. Sales of wool bring our farmers over $100 million annually.

Flax is probably our most ancient fiber crop, and it has been used throughout the centuries for making linen. At the Smithsonian, in Washington, you can see a fragment of linen that was made thousands of years ago in Egypt.

Our early American colonists practiced the art of making home-spun linen from flax; they also made "linsey-woolsey," a coarse cloth of linen and wool.

In our country today, we value flax primarily for its seed; the raw material being used for linseed oil. But fiber from flax straw is still used for linen and for various other textile products, as well as for cigarette paper.

Other vegetable fibers, mostly imported, are used for rope, twine, cordage, burlap, bags, and sacking. Among those fibers, you could include hemp, jute, abaca, henequen, and sisal. Some of them withstand contact with water; in fact, moisture may increase their strength.

COTTON IS AN IMPORTANT FIBER

Our cotton farmers produce about 12 million bales each year; they use nearly 12 million acres to grow cotton. Farmers around the world—in the West, Near East, and in the distant Orient— raise cotton. Worldwide production is about 48 million bales annually.

You will note that farmers here at home produce nearly one-fourth of the world's total. That's sufficient to supply our own needs with enough left over to make us a great cotton-exporting nation.

So if your career in agribusiness leads you into the cotton or textile industry, you may develop worldwide business contacts. The U.S. textile industry exports over $2 billion worth of textiles annually.

WHERE IS OUR COTTON GROWN?

Fourteen states comprise our Cotton Belt. These states are in the southern, south central, and southwestern regions. Texas usually leads in the number of bales produced, followed by California, Mississippi, Arkansas, and Arizona.

During recent years, our cotton production moved westward to regions where conditions favored mechanization. Cotton farms, like other farms, are getting larger and becoming highly mechanized. Mechanized cotton production requires high-capacity machinery; thus, a large capital investment. Large-acreage cotton growers may need $100,000 of working capital for their yearly operations.

HOW MUCH COTTON DO WE USE?

It takes about 48 million bales to supply world needs for a year. Here at home we use about 7 million bales annually.

Cotton Production in the United States

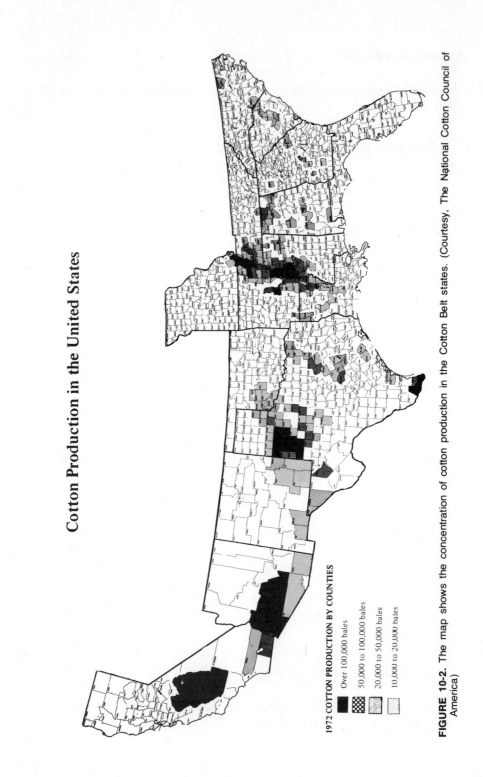

1972 COTTON PRODUCTION BY COUNTIES

- ■ Over 100,000 bales
- ▨ 50,000 to 100,000 bales
- ▤ 20,000 to 50,000 bales
- ▥ 10,000 to 20,000 bales

FIGURE 10-2. The map shows the concentration of cotton production in the Cotton Belt states. (Courtesy, The National Cotton Council of America)

Compare our use of cotton with our use of other fibers:

Per Capita Use in a Recent Year in Pounds

Cotton 13.8
Wool 0.8
Synthetic fibers 31.4

You can see that we depend on cotton for nearly one-fourth of our fiber needs. And after supplying our own requirements, we are still able to export some 5 million bales, a substantial part of our international trade.

WE USE COTTON FOR MANY PURPOSES

If you were required to list everything for which we use cotton, you would indeed have a difficult assignment. Probably no other farm product has so many uses or enters into so many different manufactured goods.

First, look at these general data on how we use all types of fibers; about two-thirds of all we use is cotton.

End-Use of All Fibers (Farm-Grown and Synthetic)[1]

	Percent
Apparel	36.8
Home furnishings	29.7
Other consumer uses	11.2
Industrial uses	17.8
Exports	4.5
	100.0

Note that apparel goods use about 40 percent of all our fibers; home furnishings, nearly 30 percent.

And our survey includes a long list of industrial uses—upholstery, auto seat covers, hose, electrical insulation, tapes, coverings, filters, rope, twine, tarpaulins, tents. Some industrial products use large quantities of cotton. In fact, cotton is our greatest industrial-use farm crop. Hose, belting, and cord for auto tires use over 16 million pounds each year.

---

[1]U.S. Census, USDA Handbook No. 524, published in 1977.

THE WORK OF THE COTTON INDUSTRY

To get cotton from producers to users is the principal function of the cotton industry. To accomplish this great task of national and global importance takes hundreds of trained specialists and industry experts. So you will find rewarding career opportunities in this worldwide activity. We will list some of them here and then describe them more fully as we trace cotton from grower to user.

Specific Opportunities in the Cotton Industry

Cotton buyer for ginner, merchant, dealer, commission person, farmer cooperative, textile manufacturer, exporter, oil seed mill, foreign purchaser, or government agency
Cotton salesperson for any of the agencies just listed
Cotton gin operator, manager, or owner
Compressor operator, manager, or owner
Warehouse operator, manager, or owner
Cotton merchant (does buying and selling)
Commission person (does buying and selling for clients)
Marketing services, including:
    "Classing"
    Compressing
Storing and warehousing
Financial services
Transportation
Insurance
Service with traders on cotton exchange
Textile manufacturer's representative
Sales promotion work with:
    Farmers' organization
    Trade association
Communications, including news reporting
Service with exporters
Government services

All of the preceding services and others aid in getting cotton from producers to users. Now let's consider some of these activities more closely.

FIGURE 10-3. After 40 years of effort, engineers produced the mechanical cotton picker. Now over 99 percent of our cotton crop is harvested by machines. (Courtesy, International Harvester Co.)

GETTING COTTON TO THE USERS

The marketing of cotton is a fascinating story. You will find that many people and many agencies take part in it. Along the marketing route, you will meet not only those who actually buy and sell cotton but also many others who render professional business services that are essential to efficient marketing.

Marketing Begins with Picking

Let's say that marketing starts when the cotton is picked. Not many years ago, almost all cotton was picked by hand; now over 99 percent is picked by machine.

The development of a machine for picking cotton is a story in itself—an example of how persistence and determination achieved what many people said "couldn't be done."

Cotton harvesting was an enormously difficult problem. For more than 40 years our engineers, several generations of engineers, struggled with the problem of the mechanical harvesting of cotton until, finally, we could announce that we had a commercially practical machine.[2]

After Picking Comes Ginning

Some 7,000 cotton gins serve the Cotton Belt—all modern offspring from that epoch-making invention of Eli Whitney in 1793. Most cotton gins serve local areas and are good examples of rural industry. They perform custom ginning for farmers, and ginners also buy, sell, and store cotton.

If you live in the Cotton Belt and want to find a business career near home, employment in a cotton gin might lead you to a

FIGURE 10-4. Ginning cotton. (Courtesy, The National Cotton Council of America)

[2]J. L. McCaffrey, formerly President, International Harvester Co.

profitable management position, or even a business of your own, with year-round activity.

Cotton gins have been substantially improved during recent years. They now have greater capacities and better facilities for cleaning and drying. The average-size gins in California and Texas now use 800 to 1,000 horsepower.

The primary function of the cotton gin is to separate the lint from the cottonseed and then compress the lint into 480-pound gin bales. For this service ginners now charge an average of about $30 per bale, including the bale ties and bagging.

Some gins also process the seed. First, they remove the "linters," short fibers that adhere to the seed. (Linters are used in making felts for mattresses and upholstery.) Then the ginners or oil mills may crush the seed and convert it into oil, cake, and meal. Cottonseed oil is used in the manufacture of shortening and margarine.

Compressed Bales for Long-Distance Shipment

The larger gins may have "compresses"—high-power presses that compress the relatively loose "gin bales" to one-half or one-third of their original size. Compressed or running bales are better adapted for long-distance shipment, and because of the lessened bulk, they can be shipped overseas much more cheaply than can gin bales.

Gin owners also may buy and sell cotton—either on a commission basis or for their own accounts. And some operate warehouses for cotton storage. Warehousing is an essential service because the producer sells his or her cotton at harvest time, but it may be weeks or even months, before the mills are ready for it.

WHERE ARE THE COTTON MARKETS?

Cotton is bought and sold at thousands of places in our country and in foreign countries. It is a commodity with a worldwide demand.

Here at home, we have various types of markets—local country markets, central markets in the larger cities, mill markets to provide cotton for "spinners" and textile makers, port markets to facilitate exporting, and our great cotton exchanges. These exchanges normally deal in "futures"—contracts for future delivery of cotton. But today our federal government supports the price of cotton at a fixed level. Hence the volume of trading in futures has drastically declined.

"Spot markets" differ from the cotton exchanges. Their function is primarily to transfer actual cotton from grower to user.

WHO ARE THE BUYERS AND THE SELLERS?

Cotton is bought and sold many times in its travel from farmer to processor. Along the marketing channels you will meet country buyers, ginners, local dealers, commission persons, cooperatives' agents, merchants' agents, mill representatives, exporters, warehouse persons, representatives of the cotton exchanges, foreign buyers, and government officials.

Anderson Clayton and Company, one of the world's largest cotton buyers, also operates cotton gins, oil seed mills, compresses, and warehouses in our country and has many plants overseas. This company, like other large buyers, may sell directly to textile manufacturers, but most of its transactions are made on the cotton exchanges.

In the cotton industry you will also meet those who provide the essential business services—bankers, insurance persons, transportation specialists, news reporters and trade journal writers, government specialists, and regulatory officials.

MARKETING SERVICES

Highly specialized services are required at various places in the marketing channels. Here are a few of them:
  1. *"Classing"*: Cotton bales from the growers come to market in mixed lots; they must be "sampled" and graded into groups of the same fiber characteristics and qualities. Accurate classing is a vital part of marketing. The different classes vary greatly in value; the best class may bring twice as much as the poorest.

FIGURE 10-5. Main cotton classing room at the U.S. Department of Agriculture in Washington, D.C. This picture shows cotton ready for classification, cotton standards boxes on tables, and stock of grade standards in background.

Cotton bales are never opened until they reach the user. To "class" them, samples must be extracted from each bale—a big job.

The U.S. Department of Agriculture keeps "official standard types" of cotton in vaults in Washington. Copies of these standards are made available for use by the experienced classers in the larger markets.

2. *Compressing:* Gins and warehouses operate high-pressure presses to reduce bale size for protection and economy in long-distance shipment.

3. *Storage and Warehousing:* Hundreds of warehouses and storage places provide the essential service of caring for the cotton until delivery is requested by the user.

4. *Financing:* Banks and other lenders supply funds to "carry" cotton from the time it is sold by the grower until it is delivered to the user. Providing these funds often requires large amounts of money and is a necessary service to those who own and "hold" cotton.

FIGURE 10-6. A professor and his class discussing yarns and fabrics. (Courtesy, American Textile Manufacturers Institute)

5. *Transportation:* Some cotton is shipped for long distances to mills or to seaports for export. Specialists study and select the best and most economical methods and routes for such shipments.

6. *Insurance:* Owners holding cotton must protect themselves against losses from fire, weather, theft, and other hazards. So insurance services must also be available.

7. *Price Protection:* Much of our cotton trade has been in "futures"—contracts for future delivery. Sellers for future delivery could protect themselves against price decline by "hedging" on the cotton exchange. Trading in futures means that when sellers contract to deliver a certain quantity at a fixed price at a definite future date, they buy, at the same time, an equal quantity to be delivered to themselves at that future date. This is the same type of hedging that is used in grain trading (Chapter 7).

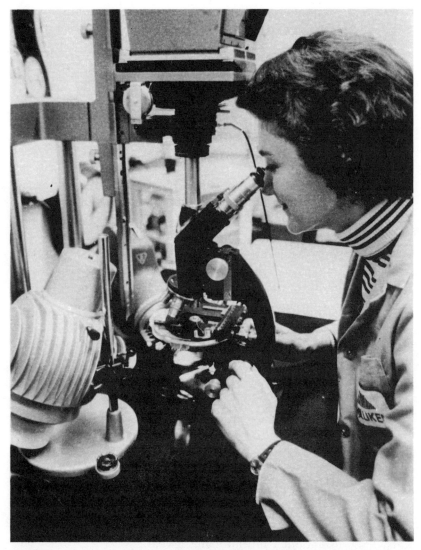

FIGURE 10-7. Using camera and microscope in studying fiber construction in the development of new fabrics. (Courtesy, American Textile Manufacturers Institute)

8. *Converting Foreign Money:* Our exporters buy cotton for dollars, but they sell in foreign countries for foreign money. So they must convert that money into dollars.

9. *Cotton Exchanges:* Our largest—in fact, for many years the world's largest—cotton exchange is in New York. England, France, West Germany, and Japan have large active exchanges which serve their trade areas and carry on national and international trade.

Members of an exchange buy and sell on order from their clients. Such customers may include domestic, shippers, cotton mills, growers, cooperatives, foreign buyers, speculators, and government agencies.

Buy and sell orders flow into an exchange from markets throughout the Cotton Belt, from central markets, from port markets, from spinners and textile manufacturers, and from clients in Europe and the Far East.

As has been stated previously, the cotton industry is a worldwide business dealing with a commodity in worldwide use.

An economist of the New York Cotton Exchange has given his appraisal of cotton marketing:

> Marketing of cotton is no child's play. It is not and cannot be an automatic or mechanized process. It involves problems that span the world and that involve millions of dollars.

The cotton industry, with all its functions, has promising opportunities for agribusiness graduates. It needs young folks who can work up to responsible positions in this global business.

COTTON MARKETS LEAD TO THE TEXTILE INDUSTRY

Some 8,000 textile manufacturers take cotton from the markets. They use about 7 million bales a year.

Textile manufacturing is a national, rather than a regional, industry. Plants are located in 44 states, although the industry is concentrated in the East and the South.

Included in the many plant operations are spinning, weaving, bleaching, dyeing, and finishing of farm-grown and synthetic fibers.

Because the cotton industry makes things that everyone needs, it is a big industry and a big employer.

> Government experts tell us that more than 15 million persons directly or indirectly derive their living from the cotton textile industry. Many millions more are linked to other parts of the textile business. And homemakers are the consumers of products from the industry in your own home town, wherever that may be.[3]

In six states of the Southeast—Alabama, Florida, Georgia, Louisiana, Mississippi, and Tennessee—the textile industry employs one out of every eight persons engaged in manufacturing.

Like other large enterprises, the textile industry requires many specialized business services. The business of selling is vast and complicated. There are many merchandising steps between the manufacturer and the user. The route is marked by activities such as creative design, fashion, promotion, advertising, packaging, transportation, financing, and collections.

The Textile Industry Looks Ahead!

> As has been true in the past and will be in the future, new horizons beckon to the textile industry. From the time of antiquity, cloth makers have striven constantly to meet the needs of users.
>
> Today you see in everyday use textile products that were unknown to your parents or grandparents.
>
> This is the result of research, the product of countless critical examinations of fibers, dyes, chemicals and mechanisms. [Research] is being continued today on a larger scale than ever before, and it is certain to grow with each passing year.
>
> You can be sure that every development in the fields of electronics and atomic energy are of interest to the industry responsible for producing fibers and fabrics.[4]

---

[3]American Textile Manufacturers Institute.
[4]*Ibid.*

EXPORT—OUR COTTON GOES
TO MANY COUNTRIES

You might think that far-distant markets are not important. But overseas sales of cotton account for much of the cash income of our Cotton Belt farmers. We usually export about 5 million bales a year, or less than half of our current production.

A large exporting concern will buy from growers or interior markets and consolidate its purchases into large lots—sometimes an entire shipload.

Our cotton exports to Canada, to the Common Market countries in Europe, and to Japan are now paid for in dollars. But much of our cotton goes to countries, other than Japan, in the Pacific area. It is sent to them under our foreign-aid programs, and we are not paid for it in dollars. Some countries pay us in their own money, which we spend within those countries.

Half the world's people live in the Pacific area and comprise a huge potential market. Our nation is driving hard to increase our exports by expanding our foreign markets and developing new ones. The U.S. Department of Agriculture works with trade groups to maintain and develop overseas markets for foods, feeds, cotton, and tobacco.

FIGURE 10-8. A room full of weaving looms ready for operation. (Courtesy, American Textile Manufacturers Institute)

Farm organizations, business groups, and commercial and banking interests are active in the drive for a greater export business.

On every continent, active cooperation between local interests and the U.S. Cotton Institute is helping to maintain cotton's position against the strong competition of man-made fibers.

We do import some cottons, mostly long staple varieties, primarily from Egypt and Mexico. But the 100,000 bales which come in each year don't seem to be very much when you compare that amount with the 12 million bales we produce each year.

OUR FEDERAL GOVERNMENT AND
THE COTTON INDUSTRY

If you want to work for the government, you may find a career in work related to cotton. At all points, from grower to consumer, representatives of our federal government have assignments in the cotton business.

Here's a brief list of some government functions:
1. Establishing cotton acreage allotments for farmers.
2. Determining the government price support rate for cotton.
3. Setting the price at which the Commodity Credit Corporation will sell cotton it has held in storage.
4. Fostering research for improvement in ginning, handling, and processing cotton.
5. Establishing export subsidy rates.
6. Supervising trading and rules of trading at the cotton exchanges.
7. Establishing standards and grades for cotton and furnishing samples to the trade for "classing."
8. Acting as custodian of farm products under the U.S. Warehouse Act—supervising the licensing of public warehouses.
9. Reporting crop and market news.

All these services require workers at many points throughout the Cotton Belt, especially at the markets, where cotton is assembled into large lots for today's mass marketing.

FIGURE 10-9. This yarn has been dipped in a computer-controlled vat. (Courtesy, American Textile Manufacturers Institute)

COMMUNICATION IMPORTANT IN
COTTON TRADING

News affecting cotton trading, developments at home and abroad must be quickly available to traders. All media of communication are employed—radio, television, telegraph, teletype, newspapers, telephones. The course of cotton trading is shaped by news. Trade journals, company publications, and financial papers all carry news about cotton—from reports of farmers' intentions to plant to those of actual consumption of cotton by users. All through the marketing route, news is being made and being reported.

Are You Interested in Agricultural Journalism?

If you are, look into the cotton industry. You will find many opportunities there. The world over, cotton probably gets more publicity and requires more writers than any other crop.

TRADE ASSOCIATIONS SERVE
THE INDUSTRY WELL

In these associations lies another kind of work that may become a goal for you. In the cotton industry, as in other agribusinesses, trade associations fulfill an important function. Agribusiness curricula at your college can help you prepare for association work.

What Kind of Work Do
Associations Do?

Trade associations serving the cotton industry provide services such as these:

Services in Government Relations: The associations make sure that the interests of the industry are well represented with the various branches of the federal government. They work with government officials on problems relating to foreign trade, taxation, labor relations, legislation, raw materials, transportation, etc.

FIGURE 10-10. This sales worker is studying fabric designs in preparation for a sales program. (Courtesy, American Textile Manufacturers Institute)

Economic Information: The associations provide their members with up-to-the-minute economic data, including statistical reports pertaining to the industry. They gather statistics and facts that managers can use in making their long-range decisions.

Publishing Periodicals: The associations prepare and distribute information on overall industry problems, legislative matters, depreciation and other tax problems, labeling requirements, and other regulations.

Committee Services: Committees are usually composed of executives, who work with the associations' staffs on such key matters as raw materials, economic policy, education, foreign trade, membership, arbitration services, taxes, traffic, and similar problems.

Services Available Through Meetings: Meetings, large and small, are conducted which are of direct interest to the industry. Annual meetings bring together a large percentage of the total membership. All the meetings enable members to exchange information and discuss industry-wide and regional problems.

SOME DESIRABLE QUALIFICATIONS FOR CAREERS IN THE COTTON INDUSTRY

College Training in Agriculture and Business: It is preferred that this be obtained at institutions in cotton-producing states.

Courses in agriculture should include agronomy, with emphasis on cotton—production, utilization, processing of fiber and seed, grading and "classing," factors determining quality, varieties.

Courses in business should bring knowledge of economics, business law, marketing, transportation, money and banking, insurance, and exporting services.

Thorough Knowledge of Cotton: This is essential. Most of the professional work deals with the product itself and is a buying-selling activity based on grades, classes, and quality standards established by experts. Without accurate product knowledge, you would be at a great disadvantage.

On-the-Job Training: You can't gain the specialized technical and professional abilities needed without "practice." In this phase of your preparation, you gain much from association with your supervisors.

Experience: Use every opportunity to broaden your practical knowledge of the industry and all of its activities from producer to user. Every time you are exposed to a new service or custom of the industry you increase in "stature." Each new experience can be a step to promotion and progress.

THE WORLDWIDE COTTON AND TEXTILE INDUSTRY OFFERS OPPORTUNITIES

Here is another prominent agribusiness. Perhaps more than any other, it is international in scope. Somewhere, in its many activities, you may find a starting place for your life's work.

SUGGESTED FURTHER READING

The American Textile Manufacturers Institute publishes an excellent set of brochures entitled *There's a Career for You in Textiles.* It is one of the best that the authors have seen. You may obtain it, if your counselor doesn't have it, by writing to the institute at this address: 1150 Seventeenth Street, N.W., Suite 1001, Washington, D.C. 20036.

The National Cotton Council of America, located at P.O. Box 12285, Memphis, Tennessee 38112, has two bulletins: "The Story of Cotton" and "Cotton from Field to Fabric." These will increase your understanding of the cotton industry. The council will send the bulletins to you at no cost.

Chapter 11

THE FARM EQUIPMENT
INDUSTRY

THE FARM EQUIPMENT
INDUSTRY

Each year we have fewer farmers. But the small remaining group gives us abundant harvests. These farmers produce more than enough for our own needs—enough so that we can send large quantities of food and fiber to other countries.

OUR FARMS ARE HIGHLY MECHANIZED

Visit some progressive farms and you will see why fewer farmers can produce more. Our farms are highly mechanized. Fast-working power machines do the field work; the farm buildings are carefully designed to save labor and to speed up the work; farmstead operations and materials handling have been "electrified," and "automated."

FARM EQUIPMENT INDUSTRY	→	FARMER, RANCHER	→	FOOD, FIBER

The farm equipment industry and the farm supplies industry (Chapter 12) have contributed much to the production efficiency of our farmers. Many others have also, especially our colleges of agriculture, our extension services, and our agricultural experiment stations with their basic science research projects.

HUNDREDS OF COMPANIES MAKE
FARM EQUIPMENT

Hundreds of companies—some in every state—manufacture the production equipment used by our farmers. The smaller com-

FIGURE 11-1. Farm operations are mechanized. Upper left—Harvesting corn with a grain combine which picks the ears from the stalks, shells the kernels from the ears, and loads the shelled corn into a truck from the delivery spout. Upper right—Accurate planting, six rows at a time. Lower—Livestock feeding with a portable feed grinder-mixer.

panies may each produce only one or two items; the larger companies make more complete lines of power machines and implements. Farm production equipment includes tractors, field machines, farm buildings, electric power units, and related items. These are more nearly permanent than the expendable supplies described in the next chapter.

Today our rural regions merge with suburban areas, so you will find some of the products of the farm equipment industry used by suburban dwellers. And some models of tractors and their attachments are used for industrial and construction work in our cities and on our highways.

THESE "GOODS" COME FROM THE FARM EQUIPMENT INDUSTRY

Here are some of the principal product groups we will include in the farm equipment industry:

> Power units, tractors, and machines
> Farm structures and barn equipment
> Electric farm equipment
> Farmstead and materials-handling equipment
> Lawn, garden, and estate equipment
> Light-industry power machines

IT IS A BIG INDUSTRY WITH GOOD CAREER OPPORTUNITIES

The farm equipment industry is big and affords promising careers. If you grew up on a farm and have agribusiness college training, you may have special qualifications. Manufacturing and merchandising today's farm equipment require working closely with farmers. The industry needs persons who can speak the farmers' language. These persons must know the farmers' crops, be aware of their production problems, and be able to recommend equipment best suited to their needs.

We will emphasize career opportunities as we describe the components of the industry—its *organization*, its *functions*, its *customs*, and its *services*. We will summarize them at the end of this chapter and mention some desirable qualifications for positions in the farm equipment industry.

THE THREE COMPONENTS

You will find that this industry, like others we have described, consists of (1) manufacturers, (2) wholesalers or branch houses, and (3) retail dealers. Most manufacturers prefer to distribute their products through local retail dealers. Retailers represent the manufacturers and are also close to the actual users. In a very real sense, they are on the front line and have a key spot in farm mechanization.

Now, let's look at each of the three components.

Manufacturers

Scattered all across our country, manufacturers produce a tremendous variety of equipment for our farmers, contractors, gardeners, and suburban homeowners.

Billions of Dollars Invested in Farm Equipment: The total number of manufactured items is amazing. Farm tractors, industrial tractors, garden tractors, power units, tillage and harvesting machines, motor trucks, electric motors and controls, pumps, water systems, conveyors and elevators, building materials, pre-fabricated farm structures—all these and many more items are produced in our factories. Manufacturers produce close to $20 billion worth each year.

Our farmers now own nearly $60 billion worth of machinery and motor vehicles.

Field Operations and Farmstead Operations Are Mechanized: As our farms grow larger they become more highly mechanized; they use bigger tractors and higher capacity implements; materials handling—a big task on the modern farm—is done with power-operated equipment. You can say that automation has been applied to the farm as well as to the factory.

You will realize this when you look at some of the new buildings on our modern farms. They are functional production units, just as are harvesting machines in the field. You will see mechanized structures. They are completely furnished with power-operated materials-handling equipment. Sanitary and well ventilated, with accurate control of temperature and humidity, they provide a favorable environment for birds and animals. Good structures enable our farmers to produce efficiently and to give us high quality products. Farmers spend over $1 billion a year for new service buildings.

This Industry Has Non-farm Customers Also: Each year more non-farm users purchase machines from the farm equipment industry. Our farm equipment makers now sell to landscaping firms, construction contractors, highway departments, parks, golf courses, rural and suburban estate owners, and home gardeners.

Visit a Farm Machinery Manufacturer: Probably there is one near you. Whether the company is large or small, you will learn much from such a visit. A visit to a major manufacturer is quite thrilling.

First, you would see the *factory.* Maybe you would start your tour in the foundry, where molten iron is poured into molds to make castings. Next might be a look at the forge shop, where heavy drop hammers forge the red-hot steel billets into desired shapes. Then you might see the machine shop, with its lathes, millers, cutters, planes, broaches, and other automatically controlled "machine tools." Here gears are cut and forgings are smoothed and surfaced to exact dimensions so they fit perfectly. (Mass production requires that identical parts be interchangeable.) You'd see many tests and inspections and the use of accurate gages for measurements. Perhaps you'd see electronic devices checking the accuracy of certain parts and automatically rejecting any sub-standard ones.

Along the way you'd see the paint shops. Some sub-assemblies and parts are dipped into large paint vats and later joined into the complete unit. Some machines are spray-painted, trimmed, and lettered after they are completely assembled.

Next you'd see the *assembly line.* Here the machine "grows" before your eyes as you walk along the line. Parts and sub-units flow onto the line, where workers assemble them into the complete product. This is a fascinating process to watch, especially when you realize that the assembly-line method is the heart of our mass-production system.

Then you might visit the *shipping department.* There the completed machines are loaded on trucks or rail cars for transport to wholesalers, to branch houses, directly to retail dealers, or to seaports for export.

But there's much more to the whole enterprise than just the factory. Products must be transported to, and sold through, branches, wholesalers, and retailers. Ultimately they must be demonstrated and sold to users, financed, and serviced through all the years of use.

Look at the Main Offices of a Large Manufacturer: You will find several major divisions and many departments within those divisions, each conducting a specialized, essential business service. Each department seeks new employees from the colleges, *and each favors agribusiness graduates.*

Here are some of the departments or divisions at a central office. Of course, actual organization varies with different manufacturers.

1. *Purchasing Department:* This section is in charge of selecting sources, procuring materials and supplies, making contracts with suppliers, and buying. A large manufacturer will have several thousand suppliers of materials, parts, and ac-

cessories. So you can safely say that "big business" is often the best customer of "small business."

2. *Sales and Marketing Division:* The job of this division includes the making of contacts with branch houses, wholesalers, and retail dealers; administration of sales contracts; advertising and sales promotion; sales planning and market research; public demonstrations and exhibits; obtaining of new retail dealers where required; improvement of dealer merchandising methods.

3. *Finance Department:* Procurement and investment of capital, credit and collections, profit planning, price research and price policies, cost control, cost accounting, and finance plans for retail dealers come under the jurisdiction of this department.

4. *Engineering Department:* Product design and improvement, field testing, production engineering, development of new machines, engineering research, and manufacturing research to find better and more economical methods of production are some of the duties carried out in this department.

5. *Consumer Relations Division:* Building customer good will and good reputation for the products; disseminating product information through all appropriate communications media; making contacts with colleges, experiment stations, and extension workers, supporting and sponsoring youth programs and scholarship policies; conducting major field demonstration and exhibits—all these make up the work of this division.

6. *Parts Department:* Although the retailer makes the final sale of repair parts, the central office has a special department for parts service, maintains stocks of parts at accessible locations, regulates parts supplies and inventory, establishes prices, processes dealers' parts orders, prepares parts order blanks and parts lists. An efficient parts department is essential to good service.

7. *Traffic Department:* The work of this department includes selection of the best and most economical methods of shipment, preparation of shipment schedules, routing and

consolidation of shipments, constant study of the rates charged by different carriers, and the search for new and lower cost transportation methods.

8. *Product Planning Division:* Looking into the future to determine what machines and products will be in demand in the years ahead; recognizing the needs for new types of equipment for the changing farm production methods; and making contacts with users, experiment stations, research agencies, and extension workers enter into the overall job of this division.

9. *Export Division:* Developing foreign markets; establishing and providing services for foreign sales representatives, agencies, or subsidiaries; making arrangements with overseas shippers, exporters, and insurance companies; and making contacts with foreign government representatives and buyers from overseas are export division functions.

   Some eight or nine of our farm equipment makers sell their products throughout the world.

10. *Executive Department:* The activities of this department include the overall control and administration of operations, policy development, and execution of policies approved by the board of directors. Many staff members provide information to top executives as bases for decision-making.

Statistics Give Us a Picture of a Major Manufacturer: International Harvester, a leading manufacturer of farm equipment, motor trucks, construction equipment, fiber, twine, and other items, provides these data for a recent year's operation.

Worldwide sales	$5.97 billion
Average number of employees	93,160
Number of suppliers in U.S. (approximate)	11,000
Manufacturing plants in U.S.	18
Manufacturing plants in other countries	30
Affiliated companies (domestic and foreign)	86
Foreign countries in which IH does business	168
Regional sales offices (agricultural)	8

Parts depots   12
Number of stockholders   128,582
Number of dealers   4,075

Two of these items, especially, are typical of companies we often refer to as part of "big business":

1. *Big business and small business are dependent on each other.*
2. *Ownership of the companies is widespread.*

You will find that these two statements apply equally to almost all our large manufacturers.

Farm Equipment Manufacturers Employ Agribusiness Graduates: Every year many agricultural graduates start their careers with farm equipment manufacturers. These companies are regular employers, and today almost all their new employees—other than factory workers—are college graduates. Very few are employed who do not have some college training, and those who lack it find the road to advancement much rougher and more difficult.

So look at college training as one of the best investments you can make.

Farm Equipment Wholesalers

Throughout our whole country, you will find manufacturers' branches, district sales offices, and independent wholesalers. Company-owned branches or district offices are the wholesale agencies of larger manufacturers; independent wholesalers serve the smaller manufacturers and the makers of specialty items. You probably know one or more farm equipment wholesalers in your community.

Why Are Wholesalers and Branch Houses Necessary? Primarily because factories are often far from the retail dealers and the users. Your dealer could hardly expect prompt delivery or quick service from a manufacturer some 2,000 miles away, even with today's rapid transportation and instant communications. Dealers want to be close to a main source of supply.

Wholesale service is important to both large and small manufacturers. It is a vital part of our outstanding distribution system; it assures the user of a ready supply of whole goods and repair parts and of prompt service. Wholesalers link the factories to the retail dealers.

Here Are Some Important Wholesale Services:
1. Assembling a relatively large inventory of equipment and parts at a central point near the retail dealers of a trade area. Wholesalers may sell the products of several manufacturers.
2. Representing manufacturers and establishing good working relations with retail dealers.
3. Securing new retail dealers where necessary; contracting with them for the sale of the manufacturers' equipment and parts.
4. Discharging the service and warranty obligations of the manufacturers.
5. Aiding retailers in sales training, sales promotion, advertising, public demonstrations, etc.
6. Assisting retailers in merchandising methods, displays, show room arrangement, etc.
7. Assisting dealers in service shop arrangement, training service mechanics, arrangement of parts departments, and parts inventory.
8. Affording management counsel to retail dealers.
9. Assisting dealers in their financing plans and programs.

Independent wholesalers handle almost all types of farm equipment—water systems, electric appliances, dairy equipment, irrigation systems, pre-fabricated farm buildings, greenhouses, lawn and garden equipment.

Some large wholesalers have their own branches, subsidiaries, or sub-agents. Certain types of equipment, such as irrigation systems, dairy installations, hay drying designs, materials-handling systems, and feedlot systems, require careful planning and layout. Wholesalers have sales engineers and qualified specialists for such sales problems. Retail dealers depend on this type of technical assistance and cooperation.

Retail Dealers

They Sell Many Items to Many Types of Customers: Our farm
equipment dealers serve a tremendous market; they sell a great
variety of products to many different types of customers. No doubt,
farm and power machinery makes up the largest part of their sales.
But dealers share in billions more of our farmers' purchases of
feed, fuel, lubricants, motor trucks, fertilizer, lime and chemicals,
building materials, and electrical appliances. Then you may add to
these the sales to industrial users, contractors, suburban estate
owners, and public agencies.

Power tools for lawns and gardens make up an important part
of dealers' sales. Each year Americans spend millions of dollars on
plant materials, garden and lawn supplies and equipment—more
than for any other hobby. They bought 5 million power lawn
mowers last year. In fact, gardening has become one of our nation's most important hobbies.

Dealers service almost all the items they sell. So well managed service shops add substantially to the dealers' incomes.

You will find retail establishments busy the year round, with
goods and services appropriate for each season of the year.

Visit a Farm Equipment Dealer: Some 17,000 retail farm
equipment dealers serve their vast market. You would learn much
by a visit to one of them. Dealers are at the end of the distribution
chain—their work and abilities govern the welfare of the whole
industry. Their recommendations for and sales of equipment that
will pay for itself in use make a profit for both seller and user, and
the whole industry benefits from the sale.

Farm equipment stores grow larger, and they are now more
highly capitalized. Their national trade publication, *Farm and
Power Equipment,* uses three groups of dealers in its "Cost of
Doing Business Study." Group A had average annual total sales of
$289,000 in a recent year; Group B, $707,000; Group C,
$1,467,000.

*A Good Place to Start a Career Is in the Farm Equipment
Industry:* Starting with a retail dealer may be a good beginning to
a career in the farm equipment industry. You could get a wealth of

FIGURE 11-2. Sales promotion demonstrations for retail dealers are arranged and conducted by the sales promotion supervisor. (Courtesy, *Farm and Power Equipment*)

basic experience. You would have a close-up view of the whole enterprise, be in on the ground floor, learn much about the farm market, and learn how experienced management runs a complex and many-sided business. You would soon have a chance to apply your agriculture-business training, prove your worth, and earn promotion.

If your dealer-employer has been successful, you can be sure he or she is a good business executive. Even though the company might be relatively "small business," it would have nearly all the problems of "big business." And you would be close enough to see how the dealer handles them.

Retailers' Problems Mean Careers for You: A retailer needs help in the following areas:
1. Retail operations—securing adequate capital for support.
2. Purchasing—carefully and judiciously purchasing and stocking items that will sell, thereby avoiding large "carry-over" at the end of the season.
3. Physical facilities—providing good physical facilities for showroom, warehouse, parts department, service shop, customer parking, office, etc.
4. Sales—hiring and training salespersons, conducting sales

promotion campaigns and public demonstrations, design-
ing and making showroom displays, using sales training
services from manufacturers and suppliers, building customer loyalty and good will.

5. Parts department—maintaining adequate inventory, training parts persons, giving prompt service, offering good
merchandising.

6. Service department—procuring and training service-
persons, selling service, establishing prices for service
jobs, building reputation for efficient service, obtaining
off-season overhaul jobs.

7. Advertising—choosing appropriate media for promoting
various lines of merchandise, for seasonal messages, and
for announcements.

8. Finance, credit, collections—financing purchases from
manufacturers and suppliers, making credit arrangements
with customers, collecting receivables.

FIGURE 11-3. Sales promotion demonstration. Retail dealers and their salespersons
witnessing demonstrations of new hay baler. (Courtesy, International Harvester
Co.)

FIGURE 11-4. "Store meeting." Employees meet to hear a discussion of future plans for this retail dealership. (Courtesy, Agway Inc.)

9. Accounting, or cost control—keeping records of accounts payable and accounts receivable, billing.
10. Pricing—establishing prices for whole goods, attachments, and parts; recognizing cost of doing business; avoiding price cutting.
11. Used equipment "trade-ins"—determining allowance for traded-in items, reconditioning and selling, selling at auctions, junking, using trade-in manuals.

Desirable Qualifications for Success in Retailing: It would be easy to add more to the preceding list, but we've included enough to show that the retail dealer must be a versatile and efficient

FIGURE 11-5. This is the type of equipment you would be demonstrating and selling to farmers while working for a retail dealer. (Courtesy, J I Case Co.)

FIGURE 11-6. Retail store manager discussing programs with his salespersons. (Courtesy, *Farm and Power Equipment*)

business administrator. He or she probably has had some years of experience and during those years has acquired and developed qualities such as these:

Good judgment
Knowledge of products
Energy, enthusiasm, and drive
Administrative ability
Technical ability
Creative sales ability
Ability to make decisions based on facts
Ability to get along with people

You too can acquire and develop these abilities so essential to successful business administration. When you buttress your agribusiness college training with "front-line" experience, you will be on your way toward executive positions.

POSITIONS WITH MANUFACTURERS AND WHOLESALERS

Manufacturers and wholesalers have most of the problems retailers have, so the same career opportunities exist in these parts of the industry also.

We can list some more careers for which agriculture-business training will prepare you:

1. *Purchasing Agent:* This is a big assignment, requiring an adequate staff. Purchases of manufacturing materials are made from hundreds or thousands of suppliers.
2. *Publications Preparation:* This includes preparing instruction manuals, repair catalogs, product information materials, advertising literature, company magazines.
3. *Territory or Zone Manager:* The chief functions are aiding and supervising retail dealers.
4. *Traffic and Transportation Manager:* This person's job involves discovering and employing the best methods and means of shipment of goods and supplies, inbound and outbound.
5. *Liaison Officer:* The major duty is making contacts with agricultural colleges, experiment stations, and extension

workers to keep your company abreast of new developments in agriculture.

6. *Market Surveying:* The purpose is to determine the potential market for a new product and the best methods of introducing, selling, and servicing the item.

7. *Product Planning:* This involves aiding in the development
of new products.

8. *Business Services:* These include office management, accounting, insurance, credit and collections, pricing, and
auditing.

9. *Export Services:* These include development of foreign
markets and service to foreign buyers and overseas shipping agencies.

IN-SERVICE TRAINING

One large machinery company, John Deere, has an excellent
in-service training program as shown here:

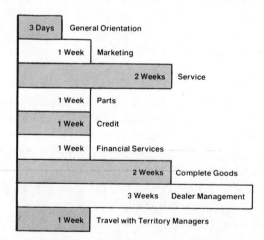

Completion of the above program could qualify you as the
company's marketing or financial services representative.

Other large equipment companies have similar programs.

QUALIFICATIONS AND PREPARATION FOR FARM EQUIPMENT POSITIONS

College training in agriculture and business is almost essential.

Courses in agriculture should stress crop production methods, principal farming enterprises, farm crops, and animal breeding.

Courses in business should include marketing, economics, salesmanship, and sales promotion, communications, and public speaking.

Knowing the language of farming—being able to talk intelligently to farmers is paramount.

On-the-job training is vital. Understanding the distribution of farm equipment from factory to farm is basic, as well as familiarity with the methods of servicing such equipment.

Experience is not the only teacher, and it may not always be "the best teacher," but it is one of the best. And almost invariably it is essential to promotion and preferment.

To provide experience and to get acquainted with possible new employees, some Massey-Ferguson dealers are offering a "Youth-in-Business" program. During the summer, a student works for a dealer demonstrating farm and home machines to customers, and locating potential buyers. If you are interested we suggest that you contact a nearby M-F dealer. It would be great experience for you.

Each year Deere and Company has a "Summer Work" program open to students who are available for one summer prior to receiving a degree, either B.S. or M.S. Deere also offers a "Cooperative Student" program with openings for freshmen and sophomore students majoring in engineering. We suggest you write to Deere and Company at John Deere Road, Moline, Illinois 61265, if you are interested in either program. Address your letter to: Personnel Representative, Corporate Recruiting.

THIS INDUSTRY IS CLOSE TO THE FARM

Perhaps it is closer to the farm than any other agribusiness. For many years it has been a regular employer of agricultural col-

FIGURE 11-7. Homeowners are customers of the farm equipment industry. Small tractors are used the year round in rural and suburban areas. Left—Keeping the driveway clear in winter. Right—Keeping the lawn in shape in summer. (Courtesy, International Harvester Co.)

lege graduates. Its representatives know the agricultural colleges and know farmers.

Lately, New Classes of Customers Have Appeared

Mechanical equipment made in our farm equipment factories is now used by golf courses, parks, suburban and rural estates, highway departments, construction contractors and builders, gardeners, and homeowners.

In the farm equipment industry, you might work in offices at headquarters in metropolitan centers, in factory cities, or in rural towns and villages. Or you might supply equipment to serve recreational enterprises near our cities and towns or furnish construction equipment for our expanding urban centers.

FARM EQUIPMENT AND A BUSINESS OF YOUR OWN

Custom and Contracting Services for Farmers

Such services may enable you to have a business of your own. You might find it a profitable and satisfactory business.

Persons Who Succeed in This Field Have These Qualifications

They know farm production problems and field operations.
They like machinery and contracting-type activities.

FIGURE 11-8. Examples of farm contracting services: Upper left—Plowing fields. Upper right—Planting a crop. Lower left—Combining grain. Lower right—Baling hay. (Courtesy, International Harvester Co.)

They have the education, training, and ability to manage an exacting business.

Here Are Some Custom and Contracting Services

Combining grain crops
Baling hay and straw
Picking and shelling corn
Picking cotton
Grinding and mixing feed
Spreading lime and fertilizer
Spraying and dusting crops for weed and pest control
Packing fruits and vegetables for growers
Operating truck pick-up service for dairy farmers
Operating farm supplies delivery service

Here Are Some Special Jobs Done by Farm Contractors

Clearing land
Leveling and reforming land surface

Constructing irrigation ditches
Installing drainage systems
Terracing and constructing water courses
Developing watersheds
Performing aerial spraying for farmers and for commer-
cial and public interests

There are many jobs to be done. Farm contracting services can be made a profitable business. A knowledge of mechanical equipment is not enough. One must maintain effective cost control through careful estimating and bidding.

Good, thorough business management in all phases of the enterprise is vital to success. You would have a chance to make practical application of all your agribusiness education.

OPPORTUNITIES WAIT FOR YOU IN THE FARM EQUIPMENT INDUSTRY

You can move from *trainee* to top management with the proper educational background. For example, Lee A. Iacocca joined the Ford Motor Company as a trainee. After serving in various field positions, he was appointed a district sales manager. Later he was appointed truck-marketing manager, and then car-marketing manager.

He took on more and more *responsibility,* was made vice-president and then president.

Mr. Iacocca is a graduate of Lehigh University and received a master's degree in mechanical engineering from Princeton University.

As well as making cars and trucks, Ford makes tractors and equipment for farmers. It is a big *agribusiness company* with opportunities for you as well as for Mr. Iacocca.

Chapter 12

THE FARM SUPPLIES
INDUSTRY

THE FARM SUPPLIES
INDUSTRY

How would you like to have a part in a $70 billion business? That's about the amount farmers spend each year for *all* the equipment, goods, and services they need to produce their crops and livestock. Today some 6 million people have jobs filling the demand of this vast market.

The farm supplies industry has a big share in the business, with annual sales approaching $35 billion.

FARM SUPPLIES CONSTITUTE ANOTHER BIG AGRIBUSINESS

Its manufacturers, wholesalers, and retailers provide expendable production supplies—fuel, lubricants, fertilizers, lime, chemicals, shipping containers, twine, wire, feed, seed, and other essentials.

These production supplies come from many different sources at home and abroad. They come from factories and fields, from mines and oil wells, and from mills, refineries, and processing plants. Finally, they reach the consumer through thousands of farm supply stores located throughout our land.

Supplies, unlike equipment, are consumed each year in the food and fiber production process.

BILLIONS ARE SPENT FOR PRODUCTION SUPPLIES

Here's what our farmers spent for certain supplies during a recent year:

Fuels, lubricants, and maintenance .... $10.1 billion
Feed and seed .... 17.0 billion
Fertilizer and lime .... 5.6 billion

We might add many other items, but the foregoing will show you that farm supplies constitute big business. The industry handles an immense variety of products, manufactured or processed by hundreds of firms.

WHO MANUFACTURES FARM SUPPLIES?

Some large corporations manufacture and distribute their products throughout the nation and to overseas markets; smaller companies serve more restricted territories.

Among the large corporations, you'll find familiar names. It may surprise you to learn that these firms have farm supplies departments: U.S. Steel, du Pont, Allied Chemical, Union Carbide, Parke-Davis, Atlee-Burpee, Standard Oil, International Minerals and Chemical, Ralston-Purina, Northrup-King, Olin Mathieson, Archer Daniels Midland, American Cyanamid.

We have named a few of the large manufacturers of farm supplies. Probably you can think of more. If you could add to these the hundreds of smaller manufacturers, you would know all the sources from which retail farm supply stores get the goods they sell, and you would know why the farm supplies industry offers attractive careers in cities, towns, and villages all across our country.

SPECIFIC CAREER OPPORTUNITIES IN THE FARM SUPPLIES INDUSTRY

Business Services

All parts of this industry, from factory to the retail store, need qualified specialists for the basic business services. These services include:

Accounting and auditing
Control of production costs, distribution costs, selling
    costs

Office management
Sales and sales promotion
Credit and collections
Communications, publicity, advertising
Customer relations

We might refer to these as conventional services. Certainly all enterprises need them. But whether you perform them in the headquarters office of a large corporation or in a retail farm store, you can give better service and avoid embarrassing mistakes if you know your company's product—know how it is used and what the benefits are that it brings to the users.

Courses in agriculture will give you those advantages.

Supplying Petroleum Products to Farmers

Manufacturers and distributors often work directly with agricultural experiment stations and cooperate with the stations in field experiments to develop new products or methods of use. Petroleum, in its many forms, powers our farm tractors, trucks, automobiles, and power units; heats nearly half of our farm homes; and provides materials for crop and animal protection.

Farmers spend $10.1 billion for petroleum products each year; farming uses more petroleum than any other single industry. All our major oil companies strive to get their share of this huge market. Each year they employ agricultural college graduates to help in this effort.

Fieldpersons: Manufacturers and distributors of farm petroleum products need qualified fieldpersons to contact dealers and farmers. These fieldpersons demonstrate methods of using petroleum products; seek to improve methods of, and equipment for, handling, storing, and applying them; develop new uses; arrange field tests and experiments to prove the value and economy of the items—especially tractor fuels, lubricants, and home-heating fuels—aid dealers with sales promotion programs.

Dealer Training: Manufacturers' representatives conduct dealer training schools to improve handling methods, accounting systems, station arrangement and services, advertising, and relations with farmer customers.

TYPICAL MODERN FRACTIONATING TOWER

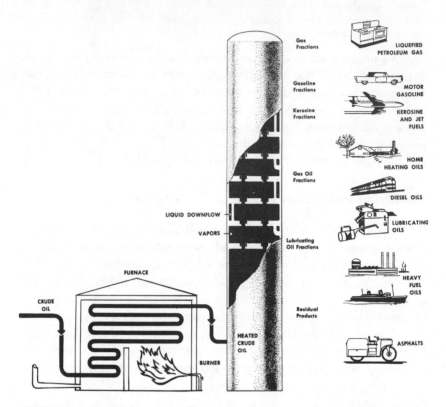

FIGURE 12-1. Fractionating tower. The fractionating process gives us all the products shown on the right. These products are used by farmers, who are the largest users of petroleum products. (Courtesy, American Petroleum Institute)

Crop and Animal Protection: Qualified specialists represent manufacturers and distributors in increasing the sales of petroleum-based insecticides and other crop and animal protection materials made from petroleum.

Supplying Fertilizer and Lime

Business Services: All parts of this supply enterprise need the "conventional" business services described earlier.

FIGURE 12-2. Nitrin at night. One of our newest nitrogen fertilizer plants, a joint venture of International Millers and Chemical Corporation and Northern Natural Gas Company. It is located at Cordova, Ill. (Courtesy, Plant Life Institute)

Manufacturers' Representatives: Special qualifications, acquired from appropriate major studies in college and from in-service training, enable manufacturers' representatives to work with chemists, botanists, and other agricultural scientists. They guide their companies in improving products and developing instructions for their use, especially in the critical requirements of accurate placement, amounts applied, time and number of applications, and methods of application.

Their duties require contacts with distributors, dealers, farmers, extension workers, and agricultural experiment stations.

Service in Manufacturing Plants: Plants are located in almost all sections of our country. More are being built to keep pace with our increasing use of fertilizers. Processes vary depending on the type of fertilizer being made. Some are relatively simple, others are quite complex.

Potential managers and administrators, especially those with education in agriculture and business, are in demand.

Services include:

> Supervising in-plant operation.
> Managing traffic and transportation.
> Scheduling manufacturing operations to meet seasonal demands.
> Taking market surveys—determining consumer preferences, optimum proportions of product ingredients.

FIGURE 12-3. United States Steel Corporation's Columbia-Geneva nitrogen fertilizer plant. Anhydrous ammonia is a gas under normal conditions. For storage and transport, however, it is compressed into a liquid which must be kept in strong pressure vessels, such as the Horton spheres shown here. (Courtesy, Plant Life Institute)

Contracting with distributors, farmer cooperatives, and large-quantity users.
Conducting sales promotion programs.
Conducting "schools" for distributors and dealers.

Services with Wholesale Distributors: Services include:

Contracting with retail dealers.
Training dealers in merchandising methods.
Preparing and distributing technical information on fertilizer use.
Publicity and advertising.
Cooperation with applicating equipment manufacturers.
Inventory control—selection of types and brands—procurement.

Field Service Contracting: Private enterprise firms and many farmer cooperatives carry on a custom service—applying fertilizer and spreading lime for their farmer customers.

Many farmers prefer this custom service, especially for applying lime to their fields. Lime application may not be required every year; hence a farmer may find it uneconomical to own a lime-spreading machine.

Such operations might give you an opportunity to engage in a field service contracting business of your own.

Farmers Spend $5.6 Billion Annually for Lime and Fertilizers: Applying lime to the fields corrects soil acidity, hastens the production of humus, and supplies calcium. Lime is not applied each year but at intervals of several years. Many farmers who do not own lime-spreading equipment employ custom operators, who treat their fields with the required amounts. Some of our farmer cooperatives, as well as private companies, have built up active lime-spreading custom services.

The manufacture and sale of commercial fertilizer have expanded greatly in recent years—a 300 percent increase within the past two decades. And farmers of many other countries use more each year. One of our manufacturers, International Minerals and Chemical Corp., now furnishes its food-producing minerals to 56 foreign countries.

Total world use will continue to increase in the years ahead. We will depend on fertilizer more and more to help produce food for the world's people.

Our farmers here at home now use nearly 12 million tons a year. Fertilizer has brought higher yields per acre and helped to reduce costs of production. Already, over 60 percent of our harvested crop land is fertilized, and more will be in the years ahead.

Several of our major oil companies are entering the fertilizer business here at home and abroad. They feel that worldwide demand will increase substantially and believe the fertilizer industry offers one of the greatest potentials in business today.

You can see fertilizer manufacturing plants and distributing centers in all our principal farming areas. Their products move out to the farms in huge quantities during planting seasons. Rapid transportation and mechanized handling are now available. Many recent improvements—higher strength concentrations, bulk delivery to farmers, and the newer liquid forms of fertilizer—have greatly reduced the labor requirement for handling and applying fertilizer.

Many farmers purchase their fertilizer material through their cooperative stores. In fact, farmer co-operatives sell about one-fourth of all commercial fertilizer, lime, and trace elements (micro-nutrients) such as boron, copper, manganese, and zinc.

AGRICULTURAL CHEMICALS— A FAST-GROWING BUSINESS

Over 50,000 chemical preparations are now registered with the government for sale to farmers. They include herbicides, fungicides, nematocides, harvest aids, animal protectants, and many others.

Chemicals are relatively new in farming. Over 80 percent of the volume of farm chemicals that farmers use today consist of material not available for use 30 years ago.

Chemicals have brought remarkable benefits to our agriculture. They do many things and help solve many problems.

Here are some things modern chemicals do for our farmers: improve and enrich the soil, help control insects, kill weeds, fumigate and sterilize the soil, treat seed, fight plant diseases, regulate time of harvesting certain crops, help livestock grow faster and use feed more efficiently, help clear brush land, eliminate rodent pests.

How Important Are Agricultural Chemicals?

This statement from Virginia Polytechnic Institute gives us this answer:

> Chemicals are essential in providing an abundance and variety of wholesome, low cost foods. Used in many ways—to nurture crops and livestock, to destroy pests and kill weeds, to cure and heal, preserve and clean— chemicals help assure the people of the United States a nutritional status as high as any in the world. Even so, each year losses of about $3 billion are caused by hordes of insect pests that chew, suck, bite, and bore away at food and feed crops and livestock. Bacteria, viruses,

FIGURE 12-4. An aerial view of Spencer Chemical Company's chemical factories on the Spring River near Pittsburg, Kans. These works, covering 168 acres, produce a variety of agricultural chemicals. (Courtesy, Plant Life Institute)

fungi, nematodes, and weeds destroy another $7 billion, making a total production loss of about $10 billion.

Without chemicals, these losses would be much greater. If we did not use these chemicals, production of many of our common fruits and vegetables would be impractical; quality of most foods would be poor. Overall, our total food supply would be drastically cut.

RETAIL SUPPLY STORE OPPORTUNITIES

How Many Farm Supply Stores Do We Have?

We have more than 36,000 according to the *Census of Business*. This figure includes farm equipment dealers, almost all of whom sell supplies. But you would have a much larger total if you counted all the smaller, local concerns—dealers in garden supplies, florist and nursery goods, etc.

What Do the Supply Stores Sell?

Here's the approximate pattern of sales by farmer cooperative stores. Sales by independently owned supply stores are quite similar:

Rank in Dollar Volume of Items Sold by Farmer Cooperatives

	Percent
Feed	37.5
Petroleum products	24.7
Fertilizer	12.8
Seed	4.7
Building materials	3.8
Farm machinery and equipment	3.3
Chemicals	1.9
Containers	1.3
Other (miscellaneous)	10.0
	100.0

Store Manager or Assistant Manager

Your education in agriculture and business makes this career a "natural." Of course, you will need to add in-service training and some experience, but that should not take long, when your basic education is so appropriate.

There are many opportunities.

Retail Store Management Brings Opportunity and Challenge: Look at the trend in retail supply stores. Such stores are getting bigger, handling more products, and engaging in more activities. They are employing more capital. And they are selling services as well as products. They are renting equipment to customers, mixing feed, doing custom milling, banking grain, custom spreading lime and fertilizer, and furnishing custom pest control services.

These stores are becoming true supply centers—large, important retail enterprises. Some centers now aid their customers in planning and conducting production programs; they offer them financial counsel and help them make credit arrangements.

As a Manager You Have to Be "on Your Toes": Look at the job

FIGURE 12-5. Typical farm supply retail store. (Courtesy, Alfred University)

of a retail fertilizer dealer. There are many thousands of such dealers in our country.

Allied Chemical Corp. wants "good fertilizer merchants." It sends this questionnaire to its distributors and dealers. Read the questionnaire carefully. It will stimulate your thinking. It applies to all farm supply stores, not just to fertilizer dealers. It will help you understand the requirements of being a good retail supply store manager.

1. Do you know your customers and potential customers well enough to understand their needs?
2. Do you handle the best products available to meet these needs?
3. Do you offer the services required to make these good products most effective?
4. Do you sell these products and services on the basis of benefits to your customers?
5. Do you follow through on your promises to be of real service to your customers?
6. Do you set your prices at a point where you can make enough money to continue to offer the best in service?

You can see that a successful dealership means more than an "over-the-counter" routine. It requires personal, social, technical,

FIGURE 12-6. Selling over-the-counter in a farm supply store. (Courtesy, Alfred University)

and professional abilities. It is not a clerk's job. It demands the best in business management and judgment. It requires a person who has formal education in agriculture and business.

Such qualities and abilities are needed at all levels of the farm supplies industry—manufacturing, wholesaling, and retailing. They are part of good management. All phases have most of the same problems and challenges. They bring you opportunity to build a successful career.

Chapter 13

CAREERS IN
ORNAMENTAL HORTICULTURE

CAREERS IN
ORNAMENTAL HORTICULTURE

A FAST-GROWING INDUSTRY

As more of our people leave the farms, our nation becomes more "urban." People flock to the suburbs. Parks, highways, public gardens, golf courses, and recreational areas increase in number. Suburban areas, adjacent to our cities, grow faster than any other. Speakers refer to our "great new suburban society."

Many of our people now have more leisure. They want more outdoor activities. Thousands find pleasure and satisfaction in gardening and flower growing; they take great pride in their lovely green lawns.

Two recent developments give us conclusive proof of the growing interest in these outdoor activities:

1. There has been a tremendous increase in the use of power lawn mowers. They are now sold by the millions, and they are sold in many different kinds of stores.
2. Country agricultural agents near cities, who used to advise only farmers, are now often swamped with requests for advice from gardeners, florists, nursery workers, and homeowners.

Other recent developments—shorter working hours, greater availability of utilities, smaller families, more consolidated schools, and increased personal income—have facilitated suburban and open-country living.

The construction of new public and private buildings, parks, golf courses, and recreation centers requires professional landscaping and ornamental materials.

These and similar changes and developments have brought

great demand and need for the products and services of ornamental horticulture. *Today it is, indeed, a flourishing business.* About 20,000 establishments produce cut flowers, flowering and foliage plants, bedding plants, and cultivated forest greens. At wholesale, their products sell for some $400 million annually. About 20,000 establishments produce nursery products, selling at wholesale for some $350 million each year.

The total sales of nearly 1,000 producers of bulb crops are over $15 million each year (wholesale value).

Wholesale value of flower seed produced annually is about $10 million.

FIGURE 13-1. The floral industry has grown to be a billion-dollar agribusiness. (Courtesy, Iowa State University)

Sales of all types of horticultural specialties by our thousands of retail stores exceed $1 billion annually.

ORNAMENTAL HORTICULTURE IS ACTIVE IN EVERY STATE

California has a multi-million dollar flower producing industry with its flowers sold through some 2,000 retail stores.

South Carolina makes "big business" from nursery products with its nearly 300 commercial nurseries.

Tyler, Texas, is the "Rose Garden of America," where rose growing is a $15 million industry employing several thousand people. The Texas Rose Festival, at Tyler, has attracted more than 100,000 visitors.

Pennsylvania, Ohio, Florida, New York, and Illinois all rank high in production of horticultural specialty crops.

Total employment by the industry has increased substantially during the last decade.

WHAT ARE THE PRODUCTS AND SERVICES OF ORNAMENTAL HORTICULTURE?

Products include:

Shrubs
Trees—evergreen and deciduous
Bulbs
Flowers
Potted plants
Sod
Topsoil
Seeds
Plant materials
Plant foods
Grasses
Ground-cover plant materials

Services include:

Landscape architecture
Garden design
Floral design

Estate maintenance
Park maintenance
Tree surgery
Construction of new lawns
Sodding
Garden maintenance
Fertilizing service
Insect and pest control
Turf management
Golf course maintenance
Landscape contracting for homes, estates, highways, and
   parks

SOME SPECIFIC POSITIONS IN ORNAMENTAL HORTICULTURE BUSINESS

Landscape Architect: This person prepares plans and garden designs for homeowners, estates, parks, and public institutions. He or she may provide supervision of landscape construction and contracting projects.

Manager of Garden Center Supply Store: Most nursery products—trees, shrubs, plants, plant foods, and equipment—are now sold by specialized garden centers. These stores also render landscaping and garden and lawn maintenance services, including fertilizing and pest control operations.

Manager or Assistant in Retail Florist Shop: This person procures flowers from the wholesale markets or from growers or shippers.

In addition to business management, the work includes styling floral arrangements appropriate for special occasions, packaging cut flowers, storing flowers, and making displays and decorations.

Park Service Worker: This person aids in the development and maintenance of public parks for a municipality or state. This is usually a civil service position.

Superintendent of Private Estate: Large estates with extensive use of ornamental plants and nursery products require specialists for adequate professional care.

FIGURE 13-2. College students studying plants in a greenhouse in preparation for careers in horticulture. (Courtesy, S. Pendrak, SUNY–Cobleskill)

Consultant: Landscape contractors, nursery workers, florists, and pest control operators need consultants.

Nursery Manager: This person's job may include responsibility for production as well as for sale of the products, in addition to general business management.

Salesperson or Sales Promoter: Employment may be found with a nursery or flower grower or in the seed industry. Qualified salespersons are in demand at all levels—by growers, shippers, wholesalers, and retail supply stores.

Manager of Packing and Shipping: Nursery products of all kinds—flowers, especially—may be shipped long distances. Growers and wholesalers require experienced specialists for these services.

Business Manager for Construction and Landscape Contracting Firm: Knowledge of landscaping requirements added to knowledge of business practices and methods is essential to this high-risk operation. Selection of basic courses in business and in ornamental horticulture will give the necessary foundation.

Supervisor of Arboretum or Botanical Garden: This person renders a service in publicly owned ornamental horticulture activities. Usually such positions are under federal, state, or municipal civil service.

Worker in Service Operations: Such services include tree surgery, fertilizing, pest control, lawn and garden maintenance, pruning, care and trimming of shrubs and hedges. These and similar services may be performed for your employer. Or you may be able to provide them as a business of your own. Often they lead to larger opportunities; many qualified persons have become proprietors of prosperous service businesses.

A MESSAGE TO YOU FROM THE AMERICAN ASSOCIATION OF NURSERYMEN

This association describes for us career opportunities and types of operations in the nursery industry. You can get further information from its office at 230 Southern Building, 15th and H Streets, N.W., Washington, D.C. 20005.

Career Opportunities

As the nursery industry continues to grow, there are excellent opportunities for additional trained personnel. Among the types of help needed are these: assistants trained in business management; wholesale and retail sales personnel; assistant nursery superintendents; general foremen (persons with background and know-how to get out in the fields and greenhouses to show others how to grow and handle the plants); propagators (those who start and develop plants to the point of transplanting); storage managers; landscape designers; and garden center managers.

From this it can be seen that the opportunities are varied and challenging. Many small nurseries urgently need ambitious young people who will become the "right arms" of the owners and grow into the businesses. Large nurseries provide a beginner with a chance to specialize in some chosen phase of the nursery business.

Nursery work encourages initiative and creative thinking, makes a substantial contribution to community bet-
terment, and offers the individual a high degree of security.

Types of Operations

Within the nursery industry there are several basic
types of business operations: wholesale, landscape, mail
order, garden center, sales yard, and agency. Most nursery firms are a combination of two or more of these basic
types and therefore offer the prospective employee a variety of career opportunities.

In the nursery business, the term "wholesale" indi-
cates the basic or original production of plants. Some
wholesale nurseries grow the plants to the "finished"
size used in landscape planting, while others grow them
a shorter period of time, selling them as small plants to a
retail or landscape nursery to grow on to the "finished"
size.

The function of a "landscape nursery" is to execute
a landscape design for the beautification of a specific
area. The designs for large areas are created by professional landscape architects. Generally small area and
home landscape designs are prepared by the landscape
nursery firm's landscape design specialist. The specialist
usually does his own estimating, selling, and supervising
of the execution of the design.

The term "mail order" refers to the specialty of advertising and selling plant materials via the mails. Some
firms specialize in producing, packing, and shipping
plants to fill the orders secured by other firms.

The term "garden center" refers to a retail business
specializing in the selling of both plants and lawn and
garden related materials such as fertilizers, pesticides,
lawn mowers, lawn and deck furniture, and other products related to garden living.

In contrast to the "garden center" is the "sales yard"
which is a display and sales area of plant materials for
retail sale.

In the nursery business there are two types of
agency operations. The first is the wholesale broker type
which acts as a wholesale selling agent for one or more
wholesale nurseries. The second is the independent
sales agent who sells plants direct to homeowners usually in conjunction with furnishing the homeowners with
planting sketches.

Since most nursery firms are a combination of two or

FIGURE 13-3. Researcher checking the growth patterns of ornamental plants. (Courtesy, Agway Inc.)

more of these types of operations, usually operated as separate departments, the industry offers career opportunities in the fields of production, marketing, and business management. Additional opportunities exist in the very important related fields of education, research and service activities, i.e., nursery inspectors, packaging specialists, product salesmen, technical advisors and others. Such opportunities may be found in both private and public employment.

THE AMERICAN ASSOCIATION OF NURSERYMEN LISTS THESE POSITIONS

Nursery worker
Nursery manager
Propagator
Storage manager
Gardener
Salesperson
General supervisor
Landscape designer
Assistant nursery superintendent
Garden center manager

Business management assistant
Landscape contractor

PREPARING FOR A CAREER
IN ORNAMENTAL HORTICULTURE

College

In college include as many as possible of the following business administration subjects:

Principles of economics
Accounting
Salesmanship and sales promotion
Communications
Business management
Marketing

Ornamental horticulture subjects should include the following:

Basics

Soils
Entomology
Botany
Plant physiology
Plant pathology

Technical and Operational Courses

Nursery management
Floriculture
Arboriculture
Plant materials
Drawing
Landscape surveying
Greenhouse management
Landscape design and construction
Nursery practices
Garden center management
Flower shop management

On-the-Job Training

Several months or even a year or more of on-the-job training may be beneficial. Under present conditions you will find it rela-

FIGURE 13-4. Sod production and lawn construction are profitable landscaping enterprises. (Courtesy, Department of Horticulture, University of Minnesota)

tively easy to find an employer who will give you a good chance to gain essential experience. And this training can be in the special field of your choice—landscape design, nursery business, retail stores, construction and contracting, service in state or municipal institutions, sales and promotion, or business management.

OPPORTUNITIES NEAR URBAN CENTERS

Consumer demand for the products and services of ornamental horticulture is great near our metropolitan areas. Today, many city-reared young men and women are planning to help answer that demand. More and more of them are enrolling in our colleges of agriculture—so many, in fact, that they make up well over half of the total enrollment. In some agricultural colleges, non-farm students make up from 60 to 90 percent of the total enrollment.

FIGURE 13-5. Greenhouse work includes checking the growth of experimental plants. (Courtesy, Agway Inc.)

And a substantial proportion of these non-farm students are studying ornamental horticulture.

There is plenty of room for them in this field, and attractive careers await them.

FURTHER READING

If the agribusiness of horticulture has a particular appeal to you, we suggest that you obtain from your school or local library the *Directory of American Horticulture*. If neither one has it, you can obtain it from the American Horticultural Society, Inc., Mount Vernon, Virginia 22121.

Chapter 14

AGRIBUSINESS CAREERS IN
GOVERNMENT SERVICE

AGRIBUSINESS CAREERS IN
GOVERNMENT SERVICE

"SERVICES" EMPLOY MORE PEOPLE
THAN PRODUCTION

Our vast agribusiness enterprise requires many special services—professional, financial, commercial, and governmental. These and similar services smooth the way and regulate and control our great *two-way* agribusiness traffic—goods from the farm move to the ultimate consumer, *and* also, goods from the industrial concerns move to the producer on the farm.

Production	Services
Manufacturing	Regulating
Producing	Controlling
Processing	Teaching
Marketing	Researching
	Reporting

Production is usually a function of private industry under the watchful eyes of governmental people at all levels.

Every year "services" become more important in our national economy. Today, nearly 54 million people—three of every five that are employed—work in services rather than in production. Perhaps the ratio is even higher in agribusiness. Certainly you will find a host of opportunities in services to agribusiness. There are about 15 million persons working in federal, state, and local government positions.

THOUSANDS OF "GOVERNMENTS" CONDUCT
AGRIBUSINESS SERVICES

You may be surprised to learn that there are so many. A large percentage of the following "governments" are responsible for some agribusiness services:

Number[1]

Federal government	1
State governments	50
County governments	3,050
Municipal governments	17,198
Special district governments	14,424

So we have about 35,000 government units of the type that may offer career opportunities for agribusiness specialists.

CAREERS IN THE U.S. DEPARTMENT OF AGRICULTURE

The U.S. Department of Agriculture offers many career opportunities, especially for agribusiness graduates. It's a big department, carrying on a multitude of activities, with some 100,000 employees; 90,000 of whom work at the 10,000 operating locations throughout the nation and in other countries. Often their work is closely allied to that of state and county agricultural workers. Some 21,000 employees work in the state extension services, and 20,000 more are attached to the state agricultural experiment stations. About 25,000 county committee persons administer the agricultural programs established by Congress.

Some Major Activities of the U.S. Department of Agriculture

Here's a condensed list of activities of this department taken from the department booklet, "You and the USDA":

> Administers, in cooperation with the states, many of the research projects so important to our agriculture.
> Provides services, regulations, and research needed to modernize, streamline, and generally grease the wheels of the whole farm marketing system.

---

[1]Statistical abstract of U.S. Census, U.S. Bureau of the Census.

FIGURE 14-1. About 10,000 department employees work here; about 90,000 work at other locations in the states. (Courtesy, Rural Electrification Administration)

Provides statistical services which inform those who make plans and decisions—farmers, bankers, businessmen, legislators, and housewives.

Administers programs to strengthen the farm economic position through income protection—not only through price programs, but also through crop insurance.

Administers credit programs to help farmers improve their farming and their living and provides a broad service to farmer cooperatives.

Carries on programs to improve the diets of Americans.

Administers extension and information programs.

Acts as the farmer's representative overseas in market development, overseas distribution programs, and in gathering information through the world-wide network of agricultural attachés.

To carry out and accomplish all these duties, thousands of workers and specialists are required. It would be difficult to describe all the specific jobs and positions within the department, but the classifications that follow will show the great variety of opportunities available.

Specific Positions and Work Areas in the
U.S. Department of Agriculture

Careers in *information and marketing* may involve the following:

> Crop and livestock reporting
> Agriculture statistics
> Market news (use of 1,500 radio stations, 170 television
>   stations, 1,600 daily newspapers, marketing services)
> Marketing service (The Agricultural Marketing Service
>   is one of the larger divisions of the department.)

Inspectors and regulatory officers are always in demand in these areas:

> Food
> Feed
> Grain
> Seed
> Dairy products and dairy processing plants
> Fertilizer
> Agricultural chemicals
> Meat and poultry (5,000 USDA inspectors are on duty in
>   meat and poultry slaughtering and processing plants)
> Fresh fruits and vegetables
> Pest control
> Plant and animal quarantine

Officers in these fields are concerned with quality control, wholesomeness, standards, grades, labeling, conditions in processing plants, and similar obligations.

Many persons are employed in the *enforcement of rules of trading*. Authority for this service is placed on the Department of Agriculture by some 30 basic laws relating to moving food and fiber from farm to consumer.

Regulation of trading and marketing is necessary at places such as these:

> Grain exchanges
> Cotton exchanges
> Stockyards
> Auction markets
> Produce markets, where perishable commodities are
>   handled. (Checks are made at shipping points and
>   terminal markets.)

FIGURE 14-2. A county extension agent, on the right, advises a farmer on a crop problem. (Courtesy, Iowa State University)

Stabilization and conservation work employs hundreds for jobs involving the following:

> Price supporting activities
> Acreage allotments
> Storage of commodities
> Sales of government-owned commodities
> Loans to farmers on stored commodities

Stabilization and conservation activities are carried on through 4 large commodity offices, 50 state offices, and 3,000 county offices. Heading this national and international service are the department's Agricultural Stabilization and Conservation (ASC) Service and its close companion the Commodity Credit Corporation (CCC).

The ASC Service deals on a day-to-day basis with hundreds of thousands of producers, carriers, exporters, handlers, warehouse workers, and others. Its operations involve the handling of millions of documents during a 12-month period.

The Commodity Credit Corporation makes loans to farmers who put surplus crops in storage.

The CCC contracts with privately owned storage agencies throughout the country. It also helps farmers, through loans, to construct their own storage and grain-drying facilities.

Stored commodities are sold by the CCC as rapidly as market conditions permit. In a recent 12-month period some $5 billion worth were disposed of by the CCC. Some were sold commercially to farmers, dealers, exporters, and others. Some were donated for distribution at home and abroad, and some were bartered for strategic and critical materials abroad.

Almost all of these activities require a working knowledge of agriculture and business practices. Agribusiness education in college is ideal preparation for beginning work in this field.

Agricultural Research Service: This is probably the largest division of the Department of Agriculture. Much of its work is in cooperation with the states, and it sends many workers overseas.

Scientists and research workers are in greatest demand. But many of the activities afford opportunity for agribusiness-trained persons also. An advanced degree is desirable.

Federal Crop Insurance Corporation: This division affords insurance protection to growers of major farm crops in about 1,000 of our counties.

DEPARTMENT OF AGRICULTURE POSITIONS UNDER CIVIL SERVICE

Most of the positions in the department are under the U.S. Civil Service Commission. To qualify you have to pass a civil service examination.

Your local postmaster can help you find where and when examinations are to be held. Or you can write directly to the U.S. Civil Service Commission, Washington, D.C. 20415. You can request "announcements" of examinations for positions in the De-partment of Agriculture.

These announcements will give the:

Date of the examination
Closing date—if any
Position to be filled
Location of work
Experience, training, and education required
Range of pay scale from lowest to highest civil service
    grade for the particular position

Also, you can contact one of the nearly 100 federal job infor-mation centers for job information for your area. If you can't find a center listed in your telephone book under "U.S. Government," dial the toll free number 1-800-555-1212 for the center nearest you and request that the information be sent to you.

Some Examples of Positions to Be Filled Through Civil Service Examinations

Agricultural commodity grader for fresh fruits, vegeta-bles, and grain
Farm credit examiner
Agricultural economist
Transportation and traffic examiner
Warehouse examiner
Food technologist
Foreign service specialist
Range conservationist
Dietitian
Cotton technologist
Agricultural extension specialist
Farm management supervisor
Forest ranger
Food service worker
Agricultural marketing specialist

Almost all such positions require college education. Some of those listed are relatively low in the pay scale of the civil service grade system. Professional workers, such as agricultural econo-mists or marketing specialists, are at the high end of the pay scale. Such positions usually require experience as well as formal educa-tion.

Federal Salary Schedule as of October 9, 1977

Grade	Example of Type of Job	Lowest Pay	Top Pay
1	Beginning file clerk	$ 6,219	$ 8,082
2	Beginning typist	7,035	9,150
3	Beginning accounting clerk	7,930	10,306
4	Senior stenographer	8,902	11,575
5	Beginning engineer	9,959	12,947
6	Mid-level computer operator	11,101	14,431
7	Mid-level analyst	12,336	16,035
8	Experienced secretary	13,662	17,757
9	Beginning attorney	15,090	19,617
10	Electronics technician	16,618	21,604
11	Experienced auditor	18,258	23,739
12	Experienced accountant	21,883	28,444
13	Personnel director	26,022	33,825
14	Experienced attorney	30,750	39,975
15	Chief chemist	36,171	47,025

Note: Due to inflation the above schedule is revised upward as the cost of living increases. This schedule is 7 percent higher than the 1976 schedule.

Note: Your counselor probably has an up-to-date schedule, or he can obtain one for you. Competition is keen for federal civil service jobs, and veterans are given priority.

OPPORTUNITIES WITH OTHER FEDERAL AGENCIES

Other federal agencies, as well as our Department of Agriculture have positions for which agriculture and business education prepares you. These agencies include:

Food and Drug Administration
Department of the Interior
National Park Service
Bureau of Public Roads
Bureau of Reclamation
Department of Commerce

Employment with all federal agencies is under civil service. Your guidance counselor will help you get information on civil service positions and the examinations for them. Your state college

FIGURE 14-3. A forest ranger works in his office and in the forests. (Courtesy, Ralph Yates)

student placement office can supply detailed information and application forms.

Currently there are about 70 applicants for each of the approximately 150,000 federal government positions available each year. The pay is good, and the benefits are excellent. Only about 10 percent of the 3 million federal workers reside in the high-cost Washington area.

• • • • • •

YOU MAY FIND A CAREER AS A STATE EMPLOYEE

Each of our states has its own department of agriculture. Take

Minnesota, for example. Its state Department of Agriculture has these basic responsibilities:

1. To encourage and develop agricultural industries (agribusiness).
2. To investigate marketing conditions affecting farm products.
3. To assist farmers, producers, and consumers in organizing and managing cooperatives and in cooperatively marketing farm products.
4. To collect, compile, and issue statistics on agricultural production.
5. To enforce laws to prevent fraud and deception in the manufacture and sale of food and other agricultural products.
6. To cooperate with the United States government in the development of the agricultural resources of the state.

The Minnesota Department of Agriculture fosters and services agribusiness and considers it highly important:

> Agribusiness—farming and related enterprises— remains Minnesota's leading industry. It accounts for 34 percent of Minnesota's employment and 27 percent of all personal income.

Principal divisions of the Minnesota Department of Agriculture are:

Administrative Services
Agricultural Products Inspections
Laboratory Services
Agronomy Services
Crop and Livestock Reporting
Plant Industries
Poultry Industries
Marketing Services

Here's Some of the Work the Department Does

Agribusiness activities of Minnesota Department of Agriculture employees include the following:

Licensing Wholesale Produce Dealers: The department acts as

FIGURE 14-4. State capitol building at Jackson, Miss., where many agribusiness persons work in the department of agriculture. (Photo, H. E. Gulvin)

a licensing agency for persons entering the business of buying and selling eggs, fruits, vegetables, hides, wool, and products of dairy plants; passes on the qualifications of applicants; and requires them to file performance bonds.

Licensing Others: This includes licensing milk and cream graders and testers, fur-farm operators, manufacturers of ice cream and frozen foods, food processing plant operations, packing house or sausage plant operations.

Auditing for Farmer Cooperatives: Auditing, accounting, and management services are provided to cooperatives. Auditors are trained in the particular problems of cooperative auditing and cooperative law.

Providing Services to Cooperative Organizations: This involves reviewing and suggesting proposals for consolidations and mergers of existing cooperative enterprises and helping with articles of incorporation, bylaws, etc.

Providing Food Inspection Services: Inspectors check all types of dairy work in plants and on farms, including the sanitary conditions of plant equipment. Inspectors also cover grocery stores, meat markets, bakeries, beverage plants, food manufacturing plants, and slaughter plants.

Providing Agronomy Services: This involves enforcing seed and weed control laws and regulations applying to the manufacture and sale of feeds, fertilizers, soil conditions, anti-freezes, insecticides, and caustic or corrosive substances.

Providing Weed Control Services: Weed infestations on private lands are controlled through working with owners; public cooperation is obtained through demonstrations, news releases, and exhibits.

Providing Feed and Fertilizer Services: Samples of feed and fertilizer are obtained from manufacturers and checked against the registration declarations. Inferior or improperly labeled items are rejected and may result in customers' obtaining refunds.

Conducting Crop and Livestock Reporting Services: Agricultural statistics are gathered, summarized, and analyzed, and the results are distributed to farmers, business persons, and others. A farm census is taken every spring by local assessors. Several hundred farmers are contacted weekly during the growing season and express their opinions as to crop production and milk and egg production on their farms. Commodity bulletins and other published reports are used by press, radio and television.

Providing Marketing Services: These services involve all marketing activities—grading, storage, transportation, packaging, advertising, communication, and price information.

Other Divisions and Functions: Insect and disease control, nursery product inspection, seed potato certification, work with the plant industry, and work with the poultry industry—all these involve certain important functions: aiding, regulating, and improving agricultural production and food products.

Each state, through its department of agriculture, carries on work like that in Minnesota. Demands on all such departments increase; almost every year they are given new duties and obligations. That means there is a good chance for your progress and promotion should you work for the state in agribusiness service.

● ● ● ● ● ●

ALMOST ALL UNITS OF GOVERNMENT NEED AGRIBUSINESS SPECIALISTS

Municipalities need such services for the overall regulation of their public produce markets; for controlling transportation of food supplies into and within the cities; for inspection of hotels, motels, restaurants, and food processing plants.

County governments need agribusiness graduates. They are directly concerned with agriculture and its products. Farm programs, established by Congress, are administered at the grass roots by county government officials.

Special government units, such as drainage and irrigation districts, watershed and soil conservation districts, need business-trained specialists as well as engineers.

AGRIBUSINESS EDUCATION LEADS TO GOVERNMENT SERVICE

Agribusiness training will prepare you for government service at the national, state, or local level. Consider opportunities carefully; get all the information you can about them. Then you can compare the advantages of government service with the advantages of employment in a private enterprise.

Chapter 15

FINANCIAL SERVICES
TO AGRIBUSINESS

FINANCIAL SERVICES
TO AGRIBUSINESS

All the industries we have described need financial services. And financial services are essential to farmers—the producers of the "merchandise of agribusiness."

LENDING MONEY TO FARMERS IS BIG BUSINESS

The business of lending money is getting bigger. Currently, outstanding loans to farmers amount to over $100 billion. The farmers' mortgage (real estate) debt is about $56 billion and the farmers' non-real estate debt is about $44 billion.

Who Loans Money to Farmers?

Here's a list of some of the principal types of lenders who provide funds for many purposes:

> Government-sponsored agencies
> Commercial banks
> Savings banks
> Life insurance companies
> Merchants and dealers
> Manufacturers
> Individuals

Our farmers' mortgage (real estate) debt was held at the start of a recent year as follows:

Million Dollars

Federal land banks 18,455
Farmers Home Administration 3,655
Life insurance companies 7,270
All operating banks 6,781
Other lending agencies and individuals 19,895

Total farm mortgage debt 56,056

EMPLOYMENT IN FINANCIAL SERVICE WOULD GIVE YOU THESE ADVANTAGES

1. You could work for the government or for private enterprise.
2. You could live in a small town, a city, or a metropolis.
3. You could make full use of your education in business and agriculture.
4. You might have both field work (travel) and office work.
5. You would have the "inside track" with agribusiness training.
6. You would be well prepared to enter this great farm credit system, which reaches from the capital in Washington into every nook and corner of our land.

SPECIFIC POSITIONS AND WORK IN FINANCIAL SERVICE

Agricultural specialist for country bank
Agricultural specialist for commercial bank (country bank, city bank, correspondent bank)
Loan appraiser for life insurance company
Secretary-treasurer or assistant for farm loan association
Secretary-treasurer or assistant for production credit association
Salesperson for farm crop insurance company
Adjuster for farm crop insurance company
Supervisor for Farmers Home Administration
Employee in credit department of manufacturer of farm equipment or farm supplies
Employee in collection department of manufacturer or supplier
Worker in credit and collection services for wholesalers and retail dealers

Representative who services loans for federal and private lending agencies

Appraiser of farm land and farming enterprises

Employee of Federal Land Bank Association

Worker in financial service with farmer cooperative (Farmers own and operate cooperatively about 9,300 businesses engaged in buying, selling, processing, and manufacturing.)

QUALIFICATIONS FOR FINANCIAL SERVICE

College Courses

College courses should include:

Agricultural economics
Accounting
Money and banking

FIGURE 15-1. Bankers discuss the merits of buying versus leasing farm equipment with a farmer. (Courtesy, Wells Fargo Bank)

Farm management
Marketing
Crop and livestock production
Farm finance
Soil management
Business law
Insurance

On-the-Job Training

This is highly desirable and is usually required by lending agencies and credit institutions. In-service training helps you gain understanding and knowledge of the methods, customs, and practices used by your employer. It will help you learn to make practical application of the academic "tools" you acquired in college and to advance to higher positions.

Experience

Credit institutions making loans to farmers usually choose employees who have farm backgrounds. Contacts with farmers through other services, such as agricultural extension services, high schools offering vocational agriculture, and experiment stations, are also given weight by employers.

Knowledge of farming and farms—land values, farm practices, production costs, and marketing potentials—is required.

Lending institutions depend on their agricultural specialists to keep their farm credit activities sound. These specialists have a heavy responsibility.

Lenders Need Specialists Trained in Agriculture and Business

The U.S. Farm Credit System has made the following remarks on careers in farm credit:

> A farm credit career requires knowledge of many interrelated fields, including agricultural economics, accounting, finance, and others. Because the field is so broad and varied, there are many ways to prepare. Much depends on you.

Of the people currently working in the system a high percentage have college degrees. Some have post-graduate degrees. Others have specialized in vocational or technical training. The system's emphasis is on finding people with the necessary knowledge and skills and the ability to learn.

Banks are modernizing their services to keep them in step with our changing agriculture and to provide adequate lending facilities for farmers and ranchers.

City banks are improving their correspondent services and their cooperation with the smaller country banks. As loans to producers become more complex, the need for trained specialists becomes urgent. Specialists are needed to improve farm record keeping and to evaluate, make, and service farm loans. Bankers and their agricultural specialists are now paying more attention to the growing importance of off-farm work by farmers.

Banks and all other farm credit agencies need representatives who know farming and farm production practices.

OUR BANKS ARE ACTIVE IN FARM CREDIT

Our bankers like farm loans. They have made many, and they want to make more.

Over one-fourth of our banks, the country over, have farm loan volumes equal to 20 percent or more of their total assets.

One-third of our banks that want more credit volume name farm loans as their first choice for new business.

Small Banks and Large City Banks Work Together

Small country banks have good working relationships—correspondent relationships—with larger city banks. Loans that are too large for a country bank are handled through participation with a correspondent. This access to larger sources of funds enables country bankers to keep pace with the increasing credit needs for larger farming operations.

Services offered by a large banking organization are indeed

FIGURE 15-2. Rural savings bank which makes loans to farmers and agribusiness companies. (Photo by H. E. Gulvin)

extensive. For example, look at this description of the facilities and services of the Wells Fargo Bank of San Francisco:

> Just tell us your problems—we'll come up with the answers. Since 1852, we have been growing with this diverse, dynamic economy—we now have more than 358 banking offices from Alturas to Holtville, from the Pacific to the Sierra. For credit information, economic facts, or comprehensive correspondent service, call on Agribusiness Administration, Wells Fargo Bank, 475 Sansome Street, San Francisco, California 94111.

Local banks are often the farmers' best help in business management. Rural banks and a growing number of urban banks help farmers with production programs, budgeting and financing, and accounting methods and offer other advisory and counseling services.

DOES FINANCIAL SERVICE INTEREST YOU?

You might be employed by one of the private-enterprise agencies we have mentioned—banks, insurance companies, individuals, manufacturers, wholesalers, or dealers. And financial service brings opportunity for government employment, especially in the vast Farm Credit System administered by the U.S. Department of Agriculture.

SOME OF THE LENDING AGENCIES OF THE UNITED STATES

Federal Land Banks: Twelve of them make loans through some 575 local associations.

Federal Intermediate Credit Banks: These banks make loans for production purposes through hundreds of production credit associations.

Banks for Cooperatives: These are regional lending institutions, owned and operated for the mutual benefit of regional and local farmer cooperatives.

Farmers Home Administration: Loans are made for the primary purpose of developing and strengthening family-type farms.

Rural Electrification Administration: Loans are made to cooperatives for construction and operation of generating plants, transmission lines, and systems to bring electric energy to rural areas.

Commodity Credit Corporation: Loans are made to farmers for crops held in storage and for construction of facilities and equipment for drying and storing crops on the farm.

A Closer Look at Federal Land Banks

The 12 federal land banks are located in:

Springfield, Massachusetts
Baltimore, Maryland
Columbia, South Carolina
Louisville, Kentucky
New Orleans, Louisiana
St. Louis, Missouri
St. Paul, Minnesota
Omaha, Nebraska

How Farmers in each of 12 Districts Share in Control of Cooperative Farm Credit System

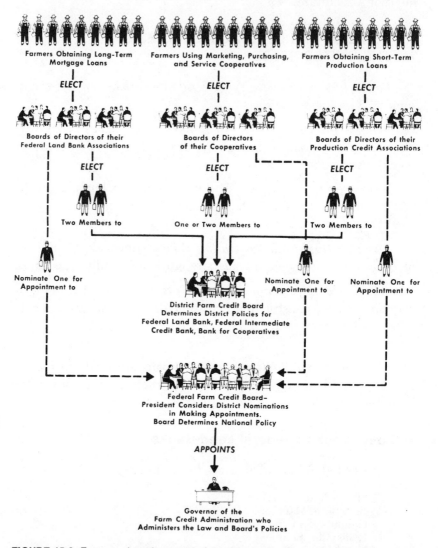

FIGURE 15-3. Farmers share in control of the Cooperative Farm Credit System. (Courtesy, Farm Credit Administration)

Wichita, Kansas
Houston, Texas
Berkeley, California
Spokane, Washington

Each federal land bank serves several of the states adjoining its location.

Although the Federal Land Bank System is now completely farmer-owned, it is given general supervision by the Farm Credit Administration of our Department of Agriculture.

Some 575 federal land bank associations, each with its own officers, employees, and board of directors, offer farmers the opportunity to apply for loans. Such an association evaluates each loan and the security for it. Approved loans are serviced by the association until paid off.

Purposes of Federal Land Bank Loans: The major purposes for which federal land bank loans are granted include:

1. To purchase or improve farm land and buildings and to construct farm buildings.
2. To purchase livestock, equipment, and supplies.
3. To pay farm operating and family living expenses, including taxes and insurance.
4. To refinance debts.
6. To provide facilities for processing, storing, and marketing farm products and handling farm equipment or supplies even if these are not to be located on the farm.

Security for Federal Land Bank Loans: Such loans must be secured by duly recorded mortgages on farm land. A loan is based on the normal value of the farm as determined by an appraiser. For further information, see Chapter 17.

Farmers Home Administration

Organization: The *national office* is in Washington, D.C.

The *National Finance Office,* in St. Louis, Missouri, handles fiscal and accounting services by keeping records and accounts for all funds disbursed and collected by the agency.

Forty-three *state offices* serve all the states, Puerto Rico, and the Virgin Islands. A state director is in charge of each state office. He and his staff direct and train county office staffs, analyze loan programs, control state budgets for loans, and approve the larger loans. The state director is assisted by an advisory committee of leading farmers and business persons.

Some 1,450 *county offices* are located throughout the country, usually in county seats. Applications for loans are made at a county office. There, the county supervisor, a capable agriculturist, helps farm families prepare farm and home plans, approves loans, gives technical advice to borrowers when he visits them on their farms, makes collections, and sends the collected funds to the National Finance Office in St. Louis.

Administration committees are part of the organization setup. Each consists of three members: two must be farmers and the third a business person who knows local farmers and their needs. Committees determine the eligibility of applicants, certify the value of farms being purchased or improved, and make recommendations concerning certain loan approval and loan servicing actions.

Purpose of Loans by Farmers Home Administration: This lending agency provides agricultural loans to worthy farm families who are unable to obtain needed credit from conventional private and cooperative lenders.

The following kinds of loans are made:

Operating loans enable borrowers to obtain farm equipment and productive livestock, pay farm operating and living expenses, make minor real estate improvements, and refinance some debts.

Farm ownership loans assist borrowers in becoming owners and operators of family-type farms. Funds may be advanced by private lenders and may be insured by the Farmers Home Administration.

Soil and water conservation loans help farmers to carry out soil and water conservation practices and to develop irrigation and farmstead water supply systems.

Farm housing loans help farmers to build and repair farm houses and other essential farm buildings.

Emergency loans assist farmers in currently designated emergency areas so they can continue farming.

Watershed loans are made to local organizations to aid in financing projects that protect and develop land and water resources in small watersheds.

GOVERNMENT LENDING AGENCIES NEED AGRIBUSINESS SPECIALISTS

To appraise, evaluate, and judge applications for any one of the loans we have just described requires a combination of agribusiness training and experience. With that combination you can find a career with one of the government lending agencies.

INSURANCE SERVICES

Selling insurance to farmers and agribusiness enterprises, servicing their policies, and adjusting their claims provide career opportunities.

Farmers need almost all the same types of insurance as urban dwellers. And they need other kinds too—especially crop insurance. That is furnished by privately owned companies and also by the Federal Crop Insurance Corporation.

Claims of loss arise from many different causes; qualified adjusters are in demand by the government and by private companies.

Most Agribusiness Enterprises Carry Insurance

Among their reasons for doing so are:
1. To protect stocks and inventory.
2. To protect stored commodities from damage.
3. To cover liability.
4. To protect goods from loss or damage in transit.
5. To protect against fire and theft losses.
6. To protect exporters against losses. (Policies insure against foreign collection hazards, political risks, currency inconvertibility, and other dangers. The Foreign Credit Insurance Association is a joint enterprise of some 70 private insurance companies and the Export-Import Bank.)

Insurance Companies—Big Investors
in Farm Mortgages

Insurance companies (non-government companies) now hold over one-fifth of our farmers' real estate (farm) mortgages. Such companies are big lenders. They need college-trained young people to prepare, through experience, to take over the serious responsibilities of appraising, evaluating, and judging farmers' eligibility for mortgage loans.

FINANCIAL SERVICE OFFERS
ATTRACTIVE CAREERS

All types of financial service agencies need well trained and capable professional workers in their activities.

A combination of business and agricultural education is a basic qualification for this special field of work.

Outlook for employment is excellent. The demand for agricultural specialists is increasing.

Give careful consideration to financial services in planning your career.

Chapter 16

OTHER AGRIBUSINESS
SERVICES AND ENTERPRISES

OTHER AGRIBUSINESS
SERVICES AND ENTERPRISES

PART A. TRADE ASSOCIATIONS

Every agribusiness has its trade association, society, or institute. It is supported by members who are active in a particular enterprise.

Here's Another Career for Agribusiness Specialists

Thousands of trade associations operate in the United States. Many of them serve some industry, and some promote certain farm products. Such organized associations, societies, or institutes are now recognized by industry and government. They have become an essential part of our business relationships.

Associations Perform Many Services

Associations render a wide variety of services to their members in areas such as public relations, promotion, communications, sales training, auditing and record keeping, publicity, transportation, research, legislative and marketing information, and many others. Most of these, and similar services, can be done better by organized groups than by individuals. So members pay dues to an association or society to receive its benefits.

Each agribusiness we have described has its own association. Some, like the farm equipment industry, have one, or more, at each level—manufacturing, wholesaling, and retailing. The food

FIGURE 16-1. National meeting of farm equipment dealers' association. (Courtesy, *Farm and Power Equipment*)

industry, of course, has many because it includes a great number of distinct enterprises—processors, canners, bakers, grocers, wholesalers, retailers, chains, supermarkets, cooperatives, and independent stores.

Some Have State and Regional Offices

In addition to their headquarter offices, some associations have offices in almost every state; and many maintain offices and staffs in Washington, D.C. A Washington office usually has the added function of keeping members informed on legislation that may affect their businesses.

A Closer Look at an Association

Let's see what an association is and what it does. Take the United Fresh Fruit and Vegetable Association, of Washington, D.C., as an example.

Here's What It Is:

> The United is an organization of shippers, growers, wholesalers, terminal market operators, brokers, some

large groups of retailers and members of allied industries. Its members handle around 75 percent of fresh fruit and vegetable marketings.

The United is unique in its field because it serves all factors in the production and marketing of fresh fruits and vegetables.

An approximate classification of United's membership is 35 percent shippers; 50 percent wholesale distributors and suppliers; and 15 percent brokers and others.

Here Are Some of the Many Things It Does: The staff keeps in close touch with legislation directly affecting the industry. Special reports are made to special groups. Wires are sent as necessary, and news of general importance is published in the association's weekly newsletter.

The association leads an industry-wide sales-building effort.

It gives distributor members a barrage of ideas and information on how to promote fresh produce.

It trains store owners, produce department managers, buyers, and produce clerks in selling and merchandising.

It helps wholesalers with problems of transportation, trucking, truck maintenance, building maintenance, materials handling, and warehousing.

The association fosters and improves trade relations among growers, shippers, and distributor organizations.

It offers members transportation services of five categories—information, legislation, carrier equipment, claim adjustments, and contacts with interstate commerce commissions.

It keeps close check on chemicals used in agriculture.

It prepares and issues publications and related items—weeklies, monthlies, yearbooks, special reports, bulletins, merchandising manuals, posters, display banners, and many types of promotional literature.

The American Soybean Association

Here's another example of an association. This one is dedicated to the promotion of a particular farm product.

These Are Its Objectives:

Bringing together all persons interested in the production, distribution and utilization of soybeans.

Collection and dissemination of information relating to both the practical and scientific phases of the problem of increased yields coupled with lower costs.

Safeguarding of production against diseases and insect pests.

Promotion of development of new varieties.

Encouragement of the interest of federal and state governments and experiment stations.

Rendering all possible services to members of the association.

Agribusiness Has Hundreds of Associations and Similar Type Groups

Some 60 groups serve the food industry. They foster and promote the interests of:

Bakers
Bottlers
Millers
Wholesalers
Retailers
Can manufacturers
Canners
Packers
Plant equipment manufacturers
Warehouse workers
Grocers
Confectioners

Other organizations promote a particular product and its use, organize research, study legislation and government policy concerning it, and aid in developing domestic and export markets. These groups and associations deal with particular products such as:

Sugar
Meat
Oil seed products
Milk
Chocolate
Corn

Cotton
Feed
Grain
Rice
Ice cream
Peanut butter

Over a hundred associations serve the farm equipment and farm supplies manufacturers and distributors. These include:

Agricultural Ammonia Institute
American Concrete Institute
American Fir Plywood Association
American Iron and Steel Institute
American Seed Trade Association
American Zinc Institute
Barn Equipment Association
Clay Products Association
Dairymen's Cooperative Association
Edison Electric Institute
Farm Equipment Manufacturers Association
Metal Building Manufacturers Association
National Farm and Power Equipment Dealers Association
National Plant Foods Institute
National Safety Council
Southern Farm Equipment Manufacturers Association
Sprinkler Irrigation Association
West Coast Lumbermen's Association

Examples of Work Assignments and Positions with Associations

The services performed by the United Fresh Fruit and Vegetable Association, mentioned earlier, will give you a good idea of the kind of work assignments in that type of organization. Of course, such assignments and special work vary with the different trade groups. But the following professional tasks are carried out by most associations.

Demonstrating and Securing Adoption of Special Accounting Systems: This is an activity that has proved very helpful to association members. Qualified personnel are employed to develop and demonstrate systems specifically designed for a particular enterprise.

Preparing Trade Literature: This may require preparation of articles for trade journals, newspaper items, general letters to the membership, business statistics, and reports.

Obtaining Market Information: This involves making predictions of the seasonal demand for a product, the probable supply, and the amount of shortage or surplus. News of new items available and of new products and processes must reach all members promptly.

Providing Merchandising Services: This includes developing sales promotional literature and materials for publicizing and popularizing a product. Also it includes taking surveys and making analyses of store layouts, displays, and employee training and sales methods, as well as making suggestions for cost control.

Conducting Market Surveys: Market surveys involve studies of population trends, changes in buying habits and consumer preferences, location of new industries within the state or community, potential demand for a proposed new product, sampling methods, and customer polls.

FIGURE 16-2. Secretary-managers of state and regional associations get together at their annual national meeting. (Courtesy, *Farm and Power Equipment*)

Providing Government Regulatory Information: The association must keep members informed of new legislation, new inspection requirements, and revised or new standards and grades for products. Information relating to labor laws, employers' liabilities, taxes and preparation of tax returns, social security payment for employees, and similar information must be supplied to members.

Conducting Meetings and Membership Drives: Trade association activities require many meetings at national, state, and local levels. The board of directors meets at stated intervals with employees of the association. Local meetings enable association employees to describe their work and explain the advantages of their service.

Associations are supported by dues from members, so efforts to increase membership are usually an obligation of association officers and employees.

Desirable Qualifications for Association Work

College training in agriculture or business is recommended. The curriculum chosen should include and emphasize these areas:

> Journalism
> Oral and written communications
> Public speaking
> Business law
> Marketing
> Economics
> Sales promotion
> Public relations
> Accounting
> Transportation

In-service training should acquaint you with the organization and operations of an association and give you a chance to meet its officers and profit by their counsel and example. Probably you will serve first as an assistant to one or more of the experienced workers.

Some associations are directly concerned with food and fiber products. Others have members that provide farmers with production equipment and supplies. Still other may carry on work that relates indirectly to farmers and farm products. But with almost all of the trade groups that serve agribusiness, your business and agricultural education will prove to be a true asset.

Agribusiness Education Paves the Way

Agribusiness education paves the way for your starting a career in trade association work. Hundreds of associations serve agribusiness. They perform many functions; they need college-trained personnel with knowledge of agriculture and business. They need persons who can write well, talk well, and meet other persons easily, and who have the energy and enthusiasm for creative, promotional effort.

• • • • • •

PART B. EXPORT-IMPORT SERVICES

Agribusiness Includes Exporting and Importing

You will find agribusiness to be an export business as well as an import business. It is a worldwide business, with buyers and suppliers in many nations, who purchase the "merchandise of agribusiness" from us and who sell some of it to us.

You might find a fascinating career in export-import service. When you complete an agribusiness college curriculum, you will be qualified to enter this field of work.

Exporting-Importing Is a Booming Business

As the world grows smaller with today's instant trans-ocean communications, world trade expands and grows larger. Nations know one another better; they buy more in foreign markets and sell more in them.

Export-import service might give you opportunity for travel and experience in foreign countries. Our manufacturers, processors, and export agencies maintain offices and branches abroad to sell their goods there and buy goods to sell here.

Each agribusiness industry you have read about in previous chapters has big export business. Our nation is the world's largest exporter of farm commodities. We send $20 billion worth abroad each year. We sell goods from our factories—tractors, motor trucks, farm machines, fertilizers, and chemicals—for use in agricultural production.

We also import about $10 billion worth of farm commodities. Primarily, these are items we don't grow at home—coffee, tea, chocolate, bananas, and tropical fruit, for example.

"Goods from the earth" and "goods from industry" flow in both directions.

> Agricultural exports and imports are a booming business, affecting our entire nation's economy. Almost every one of us is touched in some way by the buying, selling, and shipping of $30 billion worth of agricultural commodities moving in and out of our ports every year.[1]

How Important Is Exporting?

Our $20 billion export of farm products each year is about equivalent to one-fifth of the cash receipts our farmers get from their sales.

Production from 70 million of our crop acres goes to other countries.

In a recent year, farm exports required ocean shipping for 45 million tons. That is enough to fill a million freight cars or 4,500 cargo ships. The departure of 12 shiploads (average) every day of the year was required.

There was an exchange of products among more than 100 countries.

Who Buys Our Farm Exports?
Who Are Our Customers?

To the above questions perhaps you could answer, "Every country in the world." That answer would be near the truth. But, of course, some countries buy much more than others.

---

[1] U.S. Department of Agriculture.

FIGURE 16-3. Traffic department determines the best routes for export grain shipments and prepares documents for all commodity orders. (Courtesy, Bunge Corporation)

In a typical year we ship overseas farm products the amount of which is about equal to the production from the harvested acres of Minnesota, North Dakota, Michigan, and Wisconsin.

We Import Farm Products as Well as Export Them

As mentioned before, we import agricultural goods also. In fact, our country is the world's second largest importer, exceeded only by the United Kingdom.

Our *complementary* imports (commodities not produced here in quantity) include items such as bananas, coffee, rubber, cocoa, tea, spices, and cordage fiber.

But we also import *supplementary* items that add to our own production. Meat and animal products are supplementary imports.

FIGURE 16-4. From the inspection office at an export terminal, inspectors can make TV cameras zoom in for close-up views of grain being spouted into ship's holds. (Courtesy, Kansas City Board of Trade)

Only about 5 percent of the meat we eat is usually imported, although our meat imports have been increasing recently. We import some beef, mostly from Australia, New Zealand, and Argentina. Some pork is also imported, mainly from Europe. In addition, we receive large shipments of wool, mostly from Australia, the Union of South Africa, and New Zealand.

Goods from Industry Are Agribusiness Exports

Goods from industry go abroad also, and they add billions more to our agribusiness export total. Tractors, motor trucks, farm machines, fertilizer, chemicals, supplies—all go to other countries to help modernize farm production.

Some major manufacturers not only export goods but have established factories in foreign countries. We are also importing several types of tractors and farm implements, as well as a variety of farm supplies.

FIGURE 16-5. From the Cargill designed central control room, technicians monitor grain movement and handling operations throughout the complex. (Courtesy, Cargill, Inc.)

Careers in Export-Import Services

The great volume of the export-import business requires hundreds of specialists who know the techniques of business transactions and who know the merchandise of agribusiness. That's a combination you will have when you successfully complete your college study.

Think of all the professional and technical services needed in this expanding international activity. It needs efficient technical and commercial services—transportation, grading, storage, inspection, packaging. Export companies purchase commodities from our farmers, store these products, arrange for their sale to foreign countries, and provide for their shipment.

Export companies need contracts and good relations with governments, both here and abroad.

They need persons who can sell our products and help expand markets and develop new markets for them.

They need persons who are not narrow-minded and provincial but who have a broad basic outlook, who can meet persons of foreign nations and develop a live, active interest in their customs and traditions.

They need persons who know how to listen as well as how to talk.

Export services need persons who can converse with citizens of foreign lands. Consider this when planning your college program. Perhaps you will choose to learn more than one language.

Nations Strive for Greater Exports

Our nation is striving hard for greater exports. Other nations are too. Competition is keen in the race for new markets. Perhaps there is work for you in this worldwide effort.

Commercial companies, trade groups and associations, federal and state governments all join the drive to increase exports. They seek to increase sales, find new customers, and develop new markets overseas and in South America.

Offices and sales headquarters are established to advertise and demonstrate the goods and equipment offered for sale.

You'll find many trade associations with overseas branches and connections. We can't list them all, as there are so many. The U.S. Department of Agriculture works with 40 such associations in 50 different countries.

Here are a few of our many trade groups that are active in developing foreign markets:

American Cotton Manufacturers Association
American Farm Bureau Federation
American Poultry Association
Associated Tobacco Manufacturers, Inc.
National Association of Wheat Growers
U.S. Rice Millers Association

Our Export Campaign Must Inform the World

You'll read about some of its activities in your newspapers—

trade fairs, expositions, demonstrations, and product displays. You'll read about American farm products and farm equipment being exhibited in London, Tokyo, Amsterdam, cities in West Germany, Addis Ababa, and almost every sizable trade center.

Our Department of Agriculture conducts a global effort in foreign market development. It takes the lead in organizing trade fairs and expositions and arranges for foreign visitors to come to our country. And its agricultural "intelligence" service is of outstanding value. The department, through its Foreign Agricultural Service, has developed a most comprehensive news gathering and reporting system. It gets on-the-spot reports from U.S. agricultural attachés and agricultural officers at 61 points around the world. Each year nearly 5,000 reports and 2,500 foreign publications come to Washington from these overseas posts.

No agricultural industry, or any other industry, can exist without customers, customer good will, and preference for its goods. Selling is a problem common to all businesses. And it exists at all levels—manufacturing, wholesaling, retailing, and exporting.

Consider Export-Import Services in Your Career Planning

If you are thinking of a career in agribusiness, don't overlook the opportunities and challenges in export-import services. This field can enlist all of your abilities as well as the knowledge acquired in your formal training. And it might bring opportunities for foreign travel and experience, contacts with citizens of other countries and with their governments, and a broad outlook on the world of agribusiness.

Your faculty advisors and guidance counselors can help you explore this field. They can aid you in getting more information about it. And faculty members at the colleges will discuss special curricula and courses to prepare you for a career in our expanding agricultural export-import services.

● ● ● ● ● ●

PART C. GOOD COMMUNICATIONS
ARE ESSENTIAL

Every agribusiness needs good communications. It needs to tell the world about its products, needs to foster morale among its employees, needs to keep its trademark current, and needs to tell of the service it renders and its after-sale interest in customer relations.

Today agribusiness uses many communications media. Perhaps it is in this area where your interest is. Certainly you would enter a challenging and promising field of professional service.

Look over this list of communications media:

Radio
Television
Recordings
Publications
Advertisements
Exhibits

FIGURE 16-6. Student working on the layout for a newspaper. (Courtesy, University of Arizona)

FIGURE 16-7. Journalism students have many opportunities for practical experience at university radio and television stations. (Courtesy, Iowa State University)

> Motion pictures
> Magazines
> Photographs
> Farm reports
> Newspapers
> Market reports

This list isn't complete, and maybe you can add to it. But it's long enough to help you realize the many career opportunities in this vital agribusiness service. There is room for more well trained professional workers in agricultural communications.

Our colleges have special curricula that will prepare you for a career in the broad field of communications.

• • • • • •

PART D. FARM MANAGEMENT SERVICE

There is a growing demand for this highly specialized professional service. You read about farm management organizations, about the services they offer and the results they obtain. You read about banks, insurance companies, and individuals employing professional farm managers.

Perhaps this trend toward professional management exists because our farms are getting larger. They are more highly specialized and employ more capital and require larger investments. Production costs have increased markedly, and operations are more complex and intricate. So profitable operations require professional management.

Our colleges of agriculture have major curricula that will prepare you for a career in farm management. While you are in high school, your guidance counselor can help you get information on this service. And when you visit your state college, you can talk with faculty members in the farm management department. They will tell you of the opportunities, the prospects, and the challenges in this expanding professional service. They will give you examples of what farm managers do—examples like this one by Herbert N. Stapleton, formerly Manager of Shellburne Farms at Shellburne, Vermont, a highly specialized dairy enterprise:

> The scope of activity and the degree of planning and coordination required indicate that his responsibility will include physical plant; field production, methods, including machinery research and development; operation budget; and purchasing. . . . [H]is responsibility also includes finance, personnel, and livestock, pedigrees and records.

● ● ● ● ● ●

SUMMARY

Think of the changes occurring in the world today: newly emerging nations, developing nations, and the great masses of people moving toward a better life. You can sense the great advances that are coming in international trade. And you can see the changes taking place in our productive agriculture here at home.

At the base of this world progress is agribusiness. It means more food, better nutrition, and better living for countless millions.

Chapter 17

FARM COOPERATIVES

FARM COOPERATIVES

Thousands of cooperatives participate in agribusiness and within them you will find challenging career opportunities. You owe it to yourself to get acquainted with these enterprises.

Agribusiness cooperatives serve many needs and engage in a great variety of essential activities. Some cooperatives are small, serving only their local communities. Others are statewide in scope; some serve a region of several states; some are national; and some are international. Knowing the wide range of their operations, you will know why they offer so many types of careers—careers in business, industry, communications, education, public relations, research, conservation, and in dozens of other areas.

> What are farmer cooperatives?
> How are they different from other business organizations?
> How many cooperatives in agribusiness?
> What is their annual business volume?
> How can you prepare for a career in the cooperatives?

You will find answers to these and similar questions in this chapter.

HOW MANY COOPERATIVES IN AGRIBUSINESS?

Over 8,000 cooperatives market farm products and furnish our farmers with production supplies and services. Their annual business volume is about $50 billion. You will be surprised at the range and scope of their business. Here is a partial list of the work they do:

FIGURE 17-1. Two cooperative research men discuss the soil condition of the cooperative member's field. (Courtesy, Agway Inc.)

Market about one-quarter of all agricultural products.

Provide one-fifth of our farmers' production supplies.

Supply water to 25 percent of our irrigated land.

Carry one-quarter of the fire insurance on farm buildings.

Furnish electricity to millions of farm homes.

Furnish credit tailored to the farmers' needs.

Enable livestock producers to work together in associations.

Improve our dairy products through 1,500 dairy herd improvement associations.

Carry on a large share of the export and import of our farm products.

But this is far from all. In total, close to 25,000 separate cooperatives serve rural America and all of us by their participation in agribusiness.

COOPERATIVES ARE DIFFERENT

Their major objective is to furnish their member-patrons with goods and services at cost.

We have *five basic types* of business organizations in the United States.

1. *Individually owned firms*—one owner with complete control.
2. *Partnerships*—two or more owners.
3. *Corporations*—owned by investors through purchase of shares of stock; controlled by elected directors and employed managers; chartered by a state; financed by sale of stock or other debt obligations.
4. *Cooperatives*—these also are state-chartered, and most of them are incorporated. They are financed by their members through patronage, purchase of stock by members or non-member patrons, and loans from commercial banks or government agencies. Control is by the members—usually "one member, one vote." Margins above the cost of doing business are refunded to members in proportion to their patronage.
5. *Publicly owned enterprises*—municipal, county, state, and federal projects.

Operation and Management

Operation and management of cooperatives is, in many respects, similar to that of other corporations. But the objectives and goals are different. Cooperatives work primarily for the benefit of their member-patrons and seek to supply them with goods and services at cost.

Policies are developed and major decisions made by a board of directors elected by the members. This board, in turn, selects and employs the manager and, in some cases, his or her key assistant as well. Management conducts business operations and carries out the policies and decisions of the board. Management of the larger co-ops is supported by experienced business and product specialists as well as specialists in public relations, communications, advertising, agricultural science, and economics.

Clerical and operating employees are essential for the daily business routine—clerks, cashiers, fieldpersons, truck drivers, etc.

The *member-patrons* are the foundation of the cooperative enterprise. They provide financial support by their patronage and capital investment. Actual cost of goods or services to members is reduced in proportion to the success of the operation. The objective of the cooperative is to lower its costs as much as possible. Income above costs is returned to the member-patrons.

Most farmer cooperatives allow one vote per member although some types allow additional votes in proportion to a member's patronage during the preceding business year.

Control by the members is expressed through their votes at the annual meeting and at special meetings.

WHAT FARM COOPERATIVES DO

Farm cooperatives play a large part in agribusiness with an annual business volume of $50 billion. Many different types with their special functions serve the manifold needs of millions of members. Recently the Farmer Cooperative Service of the U.S. Department of Agriculture made a survey of 8,329 farmer cooperatives.

The results showed that:

> 6,568, or 78.9 percent, handled farm supplies.
> 5,842, or 70 percent, marketed farm products.
> 5,283, or 63.4 percent, performed one or more services related to marketing or farm supplies. Furnishing farm supplies, marketing farm products, and performing services for farmers are major cooperative enterprises.

Let's look more closely at these and a few others.

Farm Supply Cooperatives: Some 3,000 of them—local, regional and national—furnish farmers with the materials and equipment necessary for production. Annual sales of these supply cooperatives reach $18 billion. Products sold to farmers include fertilizers, chemicals, fuel and petroleum, automotive supplies,

FIGURE 17-2. Students are being shown how to use data processing equipment. (Courtesy, S. Pendrak, SUNY–Cobleskill)

tires, seed, feed, hardware, building supplies, paint, and almost countless other materials. At the retail level they may resemble a "general store." Other and larger cooperatives are wholesale distributing types. They carry large stocks to fill the needs of the many retail stores in their territory. Still other large concerns manufacture goods and supplies in their own factories or mills, such as fertilizer, formula feeds, and petroleum products. Cooperatives own and operate over 1,000 fertilizer plants, thousands of oil wells, and a number of refineries.

Marketing Cooperatives: More than 6,000 with an annual gross business of $24 billion handle and market one-quarter of all U.S. farm production. Measured in dollars, marketing is the largest single activity of farm cooperatives.

Many products received in a raw state from the farm are processed and prepared for consumers. Marketing cooperatives include dairy manufacturing plants, warehouses and storage depots, grain elevators, meat packing plants, and commission houses. Some marketing cooperatives specialize in handling vegetables, fruit, nuts, and similar items. Dairy products are the largest single item marketed—about one-third of the total. Then follow grain, soybeans, livestock, and poultry products.

Marketing cooperatives give the individual farmer a much larger "market place." They give him access to big volume buyers. They assemble, grade, and advertise his products, then, transport them to the most favorable market.

All this requires good management and the special skills of trained people who have both product knowledge and business acumen.

Service Cooperatives: This classification is quite broad. Actually, this type is not separate from the preceding two types because both of them also perform services for their patrons. Nearly 65 percent of those in the preceding survey performed services.

If you could see them in action, you would appreciate the wide range of their work. Here's a partial list: livestock trucking, farm delivery of feed and supplies, crop storage, crop and fruit drying, cotton ginning, custom spreading of lime and fertilizer, delivering fuel oil and petroleum products, feed grinding and mixing, livestock breeding, dairy herd improvement.

Rural Electric Cooperatives: We will classify these as service cooperatives also. You could hardly overestimate their importance. Today they bring electric power to millions of farms and to many small towns, villages, and rural residents. Rural electric cooperatives generate about one-sixth of their own power; they purchase some from federal and other generating projects, but much more from commercial power companies. And they distribute all this to their customers.

They are so important to agribusiness and afford so many career opportunities, that we shall discuss them more fully in Chapter 18, "The Rural Electrification Industry."

Irrigation Cooperatives: Farmers, especially in our western states, have joined together to form mutual irrigation companies. This movement has proved so successful that it now supplies water to one-quarter of our irrigated land.

FIGURE 17-3. Many cooperatives perform a service, such as spreading lime or fertilizer. (Courtesy, Agway Inc.)

They acquire or develop sources of water supply and distribute the water to users. Many irrigation associations require a sizable staff to carry on their critical work of allocating and distributing water fairly. Farmers in dry lands could produce little if they lacked irrigation services.

Grazing Associations: A major purpose is to enable livestock producers to use portions of our public lands under arrangement with certain government agencies. Grazing districts are established under the Department of the Interior with these objectives: "(1) Regulate occupancy and use; (2) preserve the land from abuse; (3) develop range in an orderly manner; and (4) protect and rehabilitate the grazing areas, and stabilize the livestock industry dependent on use of public land at the time the Act [Taylor Grazing Act] was passed."[1]

---

[1]Farmer Cooperative Service, USDA.

Also, grazing associations use certain areas of our national forests. "The Forest Service has sponsored more than 700 local livestock associations and advisory boards. These have played an active part in developing policies and plans for managing livestock on the national forests."[2]

Purchasing and Bargaining Cooperatives: These also render a vital service. Their objectives include:

> Saving money for their members through quantity pur-
> chases.
> Procuring the best type and quality of farm supplies.
> Providing related services.

In some types, farmers become members by purchasing a share of common stock; in others, patronage by the farmer is the only requirement. Cooperative purchasing and bargaining bring buyers and sellers together, thus enabling the individual farmer to participate in large volume sales and large volume purchases.

Bargaining and purchasing require good management and need specialists with up-to-date product knowledge and good business judgment. Bargaining cooperatives often represent their members on legal problems relating to state and federal regulations of their products.

Insurance Cooperatives: Some 2,000 farm cooperative companies furnish their members with fire insurance. Most of these companies are small—township or county—farmers' mutuals. But altogether they and some larger companies carry a large share of all fire insurance on farm buildings. Policy-holding members own the companies and control them through elected boards of directors.

National Farm Credit Cooperative Systems: This nationwide system is governed by the Farm Credit Administration in Washington. Through its many facilities and agencies, farmers can finance (1) farm ownership, (2) crop and livestock production, (3) purchase of supplies, and (4) market of their farm products. These financial privileges are made possible through four kinds of credit:

---

[2]Farmer Cooperative Service, USDA.

Long-term loans for buying land.
Intermediate credit for equipment and capital improvement.
Short-term credit for farm operating expenses.
Loans to farmer cooperatives for their development and operation.

The system is composed of three units authorized by Congress:

1. Twelve federal land banks with 575 local affiliated land bank associations.
2. Twelve federal intermediate credit banks with about 425 local affliated production credit associations.
3. Central bank for cooperatives and twelve district banks for cooperatives.

Ownership of this National Cooperative System is now entirely in the hands of farmers. And farmers also have an active part in the control of their credit cooperatives at all levels—local, district, and national.

You will find more information on financial services to farmers in Chapter 15.

Credit Unions: These also are cooperative and many of them serve participants in agribusiness. They are sponsored by a state government or by our federal government through the Bureau of Federal Credit Unions. These cooperatives seek to encourage thrift, counsel their members on intelligent use of money, and provide members with a source of low-cost loans.

Although we have well over 20,000 credit unions in the United States, fewer than 1,000 are in rural areas, but they are increasing in number.

Most credit unions belong to their own state credit union league—a type of trade association made up of the individual credit unions.

CUNA International, Inc., serves all credit unions in our states and in many foreign countries. This is a worldwide association which provides educational material and assistance.

Consumer Cooperatives: These are corporate businesses whose prime purpose is to provide a wide range of economic services on a sound business basis to their membership. They operate retail food stores, furniture stores, service stations, optical clinics, pharmacies, travel agencies, and others. Members are concerned with quality and prices.

There are many career opportunities in consumer cooperatives: Testing and rating products, performing important services in customer relations, being in charge of merchandise displays and sales promotion activities, and being head of a department.

Export-Import Cooperatives: These play an important part in our international trade in farm products. They have a substantial share in our nation's $22 billion annual export of farm products. Export business makes up to 50 percent of the total volume of

FIGURE 17-4. Sun Maid Raisin Growers of California, Kingsburg, and many other U.S. cooperatives have an important export trade—another possible career for internationally minded marketing personnel. Here Sun Maid is exhibiting its products at a Toyko trade fair. (Courtesy, Farmer Cooperative Service, USDA)

some of our cooperatives. Some are importers as well as exporters. They purchase farm production supplies abroad—grass seeds, hemp, sisal and other fibers, burlap, fish meal, molasses, twine, wire, and other items.

Here are some of our export-import cooperatives. You will find many others, some in almost every one of our states.

> Agway Inc., Syracuse, New York.
> Rice Growers Association of California, Sacramento, California
> Producers Grain Cooperative, Amarillo, Texas
> Cotton Producers Association, Atlanta, Georgia
> Rockingham Poultry Marketing Cooperative, Inc., Broadway, Virginia
> Sunkist Growers, Inc., Los Angeles, California
> California Almond Exchange, Sacramento, California
> Pacific Supply Cooperative, Portland, Oregon
> Farmers Rice Cooperative, San Francisco, California
> Caledonia Farm Seeds, Inc., Willows, California

COOPERATIVES NEED MORE MANPOWER AND WOMANPOWER

When you contemplate the vast scope of cooperative enterprises, you will appreciate their urgent need of college-trained men and women. Agribusiness leaders predict a severe shortage of qualified personnel. With our own population increasing, with world population multiplying, with per capita income and world trade rising, the demand for food, fiber, and all the services of agribusiness will grow rapidly.

Cooperatives must "gear up" for a bigger future. They seek to recruit people who can qualify for leadership.

OPPORTUNITIES IN COOPERATIVES

Would you like to seek a job with the largest farm cooperative—Farmland Industries, Inc.? If so, examine Farmland's chart presented here, and pick your college course of study:

College Curricula	Engineering and Plant Operation	Research and Development	Sales and Marketing	Administrative Management	Management Services
Agribusiness			X	X	X
Agronomy		X	X		X
Animal Science		X	X		X
Economics			X	X	X
Agricultural Engineering	X	X	X		X
Mechanical Engineering	X	X			
Petroleum Engineering	X	X			
Accounting				X	X
Business Administration			X	X	X
Finance			X	X	X
Computer Technology				X	X
Chemistry	X	X			
Mathematics		X		X	
Geology	X				
Food Technology		X			

This company also has openings in advertising, editing, and personnel work.

Careers in cooperatives are as many and as varied as their diverse activities and enterprises. It is difficult to list all the personnel needs of these thousands of business firms.

Perhaps the following general classification will help to tell you "Whom They Want" in some of the principal areas.

Operations and Business Services

Business managers and assistants
Plant and factory managers
Credit managers
Business analysts
Market analysts
Buyers—produce and supplies
Sales and service persons
Export-import specialists
Fieldpersons
Warehouse personnel
Office managers
Accountants
Cashiers and clerks
Business machine operators
Computer programmers

FIGURE 17-5. Here a class is learning accounting procedures in preparation for an agribusiness career. (Courtesy, S. Pendrak, SUNY–Cobleskill)

Professional Services

Plant and animal scientists
Poultry scientists
Agricultural economists
Home economists
Agronomists
Agricultural engineers
Soil scientists
Biochemists
Chemists
Veterinarians
Entomologists
Communications specialists

Product Specialists for:

Fertilizers and chemicals
Fruits and vegetables
Grain
Cotton
Livestock
Poultry
Irrigation

Automotive supplies and service
Rural electrification
Farm equipment
Farm supplies

Technicians for:

Dairy product quality control
Laboratories
Inspecting and testing farm products
Field work
Farm custom operations
Plant and factory operations

Finance, Credit, and Insurance Services

Credit managers
Credit union managers
Business analysts
Agricultural economists
Appraisers
Insurance co-op representatives
Actuaries
Association managers and assistants for product credit
    associations and land bank associations

Public Relations and Communication: Success of a cooperative depends in great measure upon good relations with its members, with the public, and also with its competitors. Where so
many members are co-owners and where each has equal voice in
the conduct of the business, it is vitally important that they be
kept well informed. This requires persons capable of effective
public speaking, with ability to conduct meetings efficiently, to
write well, to prepare bulletins and employ news media, and
above all, persons with the talent of getting along well with others.

Hence, people with special training and interest in public relations and communication are urgently needed. Women are well
fitted for many of these services. College training in journalism
provides good preparation for writing and editing the various publications required by our larger cooperatives.

Managers and Assistant Managers: Such positions will be the
goal of many business graduates. Cooperatives need good managers and are anxious to place promising candidates in positions
which will eventually lead them to top spots.

They need managers today and will need even more in the

FIGURE 17-6. Computer and business machines used in large regional farm cooperative. (Courtesy, Farmer Cooperative Service, USDA)

decades ahead—managers for supply stores, for marketing, purchasing, and bargaining; plant managers for feed mills, fertilizer plants, seed houses, meat packing plants, dairy and food processing plants; and farm managers.

Professionals Are Needed: The scope of the activities of cooperatives is so extensive that they need much professional talent. Included in this group are agricultural engineers, plant and animal scientists, agronomists and horticulturists, agricultural economists, home economists (more are now being employed by the larger cooperatives), chemists, soil scientists, geneticists, and entomologists.

Cooperatives conduct research in farm production, market development and business analysis, and foreign marketing opportunities. For such projects, professionally trained personnel are required.

Product Specialists: With knowledge of and interest in definite products, such specialists often reach positions of leadership.

Within this group we can include those whose training and experience give them a valuable background in specific areas such as grain, cotton, tobacco, fruits and vegetables, livestock and meat, poultry and dairy products, electric utility industry, fertilizers and agricultural chemicals, farm machinery and equipment, and irrigation.

In many cases, students can shape their college studies, with the aid of faculty counsel, to put extra emphasis on one or more of these areas of agribusiness.

Business Services: Due to the many different cooperative enterprises, a great variety of services are needed affording numerous career opportunities for accountants, business and market analysts, sales and service personnel, appraisers, field representatives, publicists, office managers, business machine operators, personnel administrators, cashiers, clerks, and secretaries.

Finance, Insurance, and Credit: These cooperatives offer opportunities for actuaries, business analysts, economists, insurance salespersons, and claim adjusters. Well qualified employees may become managers of production credit associations, credit unions, land bank associations, and other agencies serving financial needs of farms.

Farmers' capital needs will increase as farms grow larger and the need of equipment and production supplies becomes greater. Even now, some farmers count their annual credit needs in hundreds of thousands of dollars.

Laboratory and Field Technicians: These technicians are required in certain service cooperative activities. Some of these positions can be filled satisfactorily by vocationally trained people. Their work includes dairy product quality control, grading and inspection of farm products, landscape and nursery technology, laboratory technology, grain technology, seed technology, inspection in processing plants, and farm operations such as spraying, spreading fertilizer and lime, combining grain, baling hay, etc.

Export-Import Service: The increasing volume of foreign trade of cooperatives requires more specialists in the field. Major emphasis on foreign trade and commerce during college courses will help candidates qualify for responsible positions.

PREPARING FOR A CAREER IN COOPERATIVES

Some agricultural colleges and some colleges of business administration offer special studies of cooperative organization. Take advantage of all such course work if you look forward to a career in cooperatives. Your faculty advisor, other faculty members, and department heads can help you shape a suitable curriculum.

The information in Chapter 4 applies to careers in cooperatives as well as to other agribusiness concerns. You may find it necessary and desirable to have some courses in the college of agriculture and some in the college of business administration.

Starting salaries in cooperatives are equal to those in other

FIGURE 17-7. Instructor of an in-service training class preparing new employees for engineering work with a rural electric cooperative. (Courtesy, Rural Electrification Administration)

business concerns. Your starting salary will depend upon the position assigned to you and also upon your formal education. Advanced degrees bring higher salaries and enable one to start "higher up on the ladder."

Graduate study is highly desirable and you will recover its cost many times over during your working life. Go as far as you can. Cooperatives and agribusiness can use all the advantages brought by persons with higher education.

Opportunities for advancement are good. In-service training is offered in most cooperative enterprises. Frequently you can hear the manager of a cooperative report that most of the positions in the operation provide training that prepares employees for advancement.

FURTHER READING

The Farmer Cooperative Service, USDA, Washington D.C. 20250, has a circular (#33), "Cooperatives in Agribusiness." This contains a very thorough explanation of all the aspects of cooperatives. You may obtain a free copy by writing for it.

The Cooperative League of the USA has a good pamphlet called "Careers in Cooperatives." You may obtain a free copy by writing to the League at 1828 L Street N.W., Washington, D.C. 20036.

Chapter 18

THE RURAL
ELECTRIFICATION INDUSTRY

THE RURAL
ELECTRIFICATION INDUSTRY

You may find that rural e ctrification offers more career opportunities than other segme s of agribusiness and agricultural
industry. We will describe these opportunities in this chapter and
the work of the agencies and enterprises of rural electrification.
They all seek the service of graduates in *agriculture, home economics, business administration,* and *engineering.*

But first, let's get a little background information on our entire
electric industry and then look at rural electrification.

Electricity, now widely used but still a mysterious force,
serves each of us every day. What would we do without it? Surely,
chaos would result if we were deprived of it suddenly and completely. Some parts of our country know the confusing, even ter-
rifying, results from a long "blackout." We don't realize how much
we depend upon electricity until on rare occasions the "current
goes off." Almost beyond count are our daily needs that are served
faithfully and dependably by electricity.

Yet, in its practical, usable form it is a fairly recent addition to
our economy. We owe its *discovery* to the imagination of ancient
dreamers and we owe its *development* to the many scientists and
inventors who followed the dreamers. These early dreamers wondered about the lodestone that attracted iron; wondered about the
"amusing" results when they rubbed amber with a cloth; wondered whether there was any relationship between these two
phenomena. Then came the scientists. They not only wondered,
but experimented. Gilbert produced electricity by friction and
gave us that new word, "electricity." Galvani and Volta gave us
the electric battery; Coulomb and Ampere found how to measure
magnetism and electricity; Oersted found that a magnet could be

used to produce electricity; Faraday, scientist-inventor, made the first electric motor in 1821 to be followed shortly by the dynamo generator. Then in 1878, Edison used direct current from his dynamo to illuminate his carbon filament lamp.

Both power and light had come because of the relationship of magnetism and electricity.

Today, 10 decades after Edison, our nation is covered with a vast network of transmission and distribution lines that carry electricity to all of us wherever we may be. And these "life lines" are fed by large, interconnected generating systems; one can take over for another in case of need. Service is reliable; power failures are rare.

This is our fastest growing industry, doubling in the production of electric power about every 18 years. Indeed, it is a growth industry. With our increasing population, and with a multitude of new uses for electricity, it will grow even faster in the years ahead—about 200,000 new customers each year.

WHO MAKES OUR ELECTRICITY?

Who is in the business of *generating* our electric power? Here's the way this business was shared in a recent year:

Producer	Amount
Investor-owned (commercial) utilities	76.4 percent of total
Federal generating stations	13.6 percent of total
Municipal- and state-owned stations	9.3 percent of total
Cooperatives	0.7 percent of total

WHO DISTRIBUTES OUR ELECTRICITY?

Whom do we consumers buy our electricity from? In a recent year we bought:

78.8 percent of our needs from investor-owned utilities.
13.4 percent from municipal- and state-owned utilities.
7.8 percent from cooperatives (which buy current to distribute to their patrons).
0.0 percent from federally owned stations.

WHAT ENERGY FUELS THE GENERATORS?

Today, nuclear energy, with its fascinating possibilities, receives much publicity. In years to come it may supplant the fuels we now burn to make the steam that drives our generator turbines. And some day we may economically produce electricity directly from nuclear reaction. In 10 years the use of nuclear-produced power has increased six fold. In one state, Connecticut, 58 percent of the electricity used is from nuclear power.

> 46.6 percent generated by coal
> 14.4 percent from gas
> 10.4 percent by water power
> 16.8 percent from oil
> 11.8 percent by nuclear energy

FIGURE 18-1. Connecticut Yankee Nuclear Plant producing electric power for the New England area. (Courtesy, Northeast Utilities)

Throughout most of our nation, and from imports, suitable energy sources (fuels) are available. With these fuels, various private and government agencies "make" our electricity and transmit and distribute it to us. All America is now electrified. But only recently did folks in our rural areas join the rest of us who depend upon electricity.

In June 1977, the Federal Energy Administration produced the figures below for the energy sources of electricity then and in 1995. There will be definite need for electrical engineers as expansion proceeds, especially those trained in nuclear plant operation.

Total U.S. Electrical Generating Capacity

Energy Source	Existing	Projected	Total
	----------------(percent)----------------		
Coal	38	38	38
Oil	25	6	17
Gas	14	0	8
Hydro	12	7	10
Nuclear	9	45	24
Other	2	4	3
	--------------(megawatts)--------------		
Energy Output	545,364	390,594	935,958

THE PROGRESS OF RURAL ELECTRIFICATION

In 1886, Stephenson showed how alternating current might be transmitted economically over long distances. Then rural electrification became possible. In theory, at least, electric power could then serve farms and open country as well as the cities. (Today, high voltage direct current is used for long-distance transmission.)

But many difficult problems had to be solved. Serious obstacles stood in the way of progress. Many said, "It can't be done." Much pioneering work was needed—research, experimental work, demonstrations, consumer education, and community organization.

> Open country conditions were different from those in the cities.

Potential rural customers were few and widely sepa-
rated.
Farmers wanted lights only.
Lights alone would not be a profitable load for power
companies.
Farmers saw few uses for electric power.
Open-country electric lines were costly—up to $2,000
per mile.
Who would pay for such costly lines?
Generating stations were small, without capacity to
reach out.

Despite all such obstacles, enthusiasm for rural electrification
mounted year by year. The time had come! Farmers had shown
their desire for electricity through the thousands of individual
gas-engine and wind-driven lighting plants scattered throughout
the country. These were not adequate, but they did reveal some of
the important benefits of electricity on our farms.

Then powerful forces with their dedicated workers joined the
movement:

U.S. Department of Agriculture
State agricultural colleges
State engineering colleges
Electric utility companies
Farm leaders
National Grange
American Farm Bureau Federation
National Electric Light Association

These were followed by the:

Edison Electric Institute
Committee on the Relation of Electricity to Agriculture
American Society of Agricultural Engineers

Researchers demonstrated many ways in which electric power
could be used profitably in farm operations. Well over 200 such
uses had been demonstrated by 1925. And in that year, nearly 60
commercial utility companies had organized rural service departments. By 1935, 10 percent of our nation's farms were "electrified."

FIGURE 18-2. A fully electrified farm near Sycamore, Ill. (Courtesy, USDA Photo Service)

Then Came REA!

Established by Congress in 1936, the Rural Electrification Administration (REA), within our Department of Agriculture, provided low-interest, 35-year loans for bringing electricity to our farms and rural areas.

REA makes loans to cooperatives for:

> Financing the construction and operation of generating plants, transmission lines, and distribution lines for furnishing electric energy to persons in *rural areas* who are not receiving central station services.
>
> Financing home wiring and the purchase of electric appliances, equipment, and plumbing.
>
> Financing telephone service in rural areas.

Joint action by the investor-owned power companies and the rural electric cooperatives brought rapid progress. Now, nearly all farms are electrified.

RURAL ELECTRIC COOPERATIVES

Today these number nearly 1,100. They operate in most of our 3,072 counties. They are organized much the same as the other cooperatives we have discussed in Chapter 17. Each is an independent, locally managed enterprise. It is owned by its member patrons who pay a membership fee of $5 to $10. Because each local has an exclusive service right in its specified area, anyone who wants service must become a member. Income over cost is refunded to members as a cash dividend or in lower rates.

Our rural electric cooperatives serve about 8 percent of the nation's electricity consumers. They distribute about 4 percent of all electric power, operate about one-half of the transmission lines in serving their 5,000,000 members, and service over half of our farms.

Each local cooperative can obtain capital funds from REA and also engineering aid. Management and accounting assistance are also available from a nearby REA fieldperson or accountant.

A typical rural electric cooperative:

> Serves some 6,000 customers (the largest serves 35,000).
> Employs about 30 persons.
> Operates approximately 1,500 miles of transmission lines.
> Has a modern headquarters building which is usually a center of business activity.
> Furnishes the average farmer customer about 750 kilowatt hours per month.
> Has an annual revenue of approximately $1 million.
> Buys much of the power it distributes from commercial or public sources. (Nationally, the cooperatives generate about 20 percent of the power they distribute.)
> Obtains much of its revenue from non-farm customers.

Each of our 1,100 rural electric cooperatives needs qualified personnel, such as business managers, electrical and agricultural engineers, customer relations specialists, farm and home demonstration agents, community organizers, equipment salespersons, farm product specialists, accountants, and office managers.

Such cooperatives offer hundreds of attractive careers for college graduates.

COMMERCIAL POWER COMPANIES AND RURAL ELECTRIFICATION

Investor-owned utility companies generate over three-fourths of our total electric power and supply it to about the same proportion of our nation's users including about half of our farms. These companies were pioneers in rural electrification. In fact, some "historians" say it was born in 1898, when a California farmer

FIGURE 18-3. In-service training of new employees of a rural telephone cooperative. (Courtesy, Rural Electrification Administration)

used a 5-horsepower electric motor to operate his irrigation pumps.

The nationwide interest of the utility companies was well demonstrated by their early (1921) action in sponsoring, through the National Electric Light Association, the Committee on the Relation of Electricity to Agriculture (CREA). CREA accomplished much of the basic research and experimental work essential to solving rural electrification problems. Also, it brought about the coordinated efforts of agencies and organizations that were directly concerned with farming. Then CREA aided in the development of equipment and appliances that could be used profitably for hundreds of farm jobs that formerly required hand labor.

Soon, most of the power companies organized and expanded their rural service departments. Today, these departments bring valuable services to farmers, rural residents, business concerns, and industries. Their well-trained, experienced rural service managers and farm advisors have improved farm practices and made them more profitable. And by introducing and servicing new rural industries, they have brought new economic life to whole communities.

Power Companies Help Build Communities

Here is an example, reported by Edison Electric Institute:

> An agricultural area in East Texas was in economic straits in the early 1960's. Main problems were soil depletion and general loss of farm people to the cities.
>
> To strengthen the local economy, an investor-owned power company launched several broad programs. "B.I.G." (Blackland Income Growth) aimed at better soil management in a black clay area. Another program, "B.E.T." (Build East Texas), dealt with problems in a sandy belt.
>
> Together these programs sought to introduce and establish better farm technology, better fertilization and soil management, more mechanization, increasing cooperation between businessmen, farmers, and ranchers.
>
> Related activities included upgrading the appearance of buildings and grounds in small communities, farms, and ranches. This, together with other community development efforts, attracted 269 new industries, many agriculturally oriented, to the power company's service area. These additions brought more than 19,000 new jobs and an annual payroll of some $80 million into the region. For the 36 counties involved, farm income increased $51 million in the first two years of the programs.

Another Type of Community-Building

Power companies harness our rivers to make electricity for our farms, factories, and cities. Try to visit a hydroelectric power development. It is a thrilling sight. Probably you have one in your own state.

In California, the swiftly rushing water from the snow-capped Sierras is halted many times on its way to the ocean. Blocked by dams, it often lies quietly in reservoirs, ready and available for many uses.

Were you to look down from a great height, you would see not one but many dams and reservoirs along the river's descending

FIGURE 18-4. Power companies harness our rivers. Water from the high Sierras gener-
ates power for farms, factories, and cities of central California. (Photo by Archie A.
Stone)

course, resembling a giant staircase. Each succeeding step or re-
servoir provides water to drive great turbine generators, water for
irrigation and domestic use, a lake for boating and fishing, and
storage space for flood control. The high-mountain water is used
and reused many times before it can rest in the sea. Along one
California river, Pacific Gas and Electric Company has more than
a dozen dams and reservoirs.

With abundant water and electric power, the great central valley of California has become outstanding in our nation's agriculture. With an almost endless variety of high-value field crops, fruits, vegetables, horticultural specialties, and highly developed poultry and livestock industries, it is a major producer of the "merchandise of agribusiness." Upon leaving the valley, this merchandise passes through the channels of agribusiness to reach all our states and many foreign countries.

But water power alone cannot even begin to supply all the needs of the farms, industries, and cities of northern and central California. To do so, Pacific Gas and Electric Company has developed a vast interconnected system. All forms of energy—hydro, steam, atomic, gas, geothermal—are used. Today this big power company has 76 generating stations with 96,000 miles of power lines and more are being built.

Improving the Environment

Many electric power companies are interested and active in improving the areas in which they work. For example, the Northeast Utilities, a shareholder-owned power company in New England, has sponsored boating, hiking, camping, fishing, swimming, and picnicking in the Northfield Mountain area of western Massachusetts. Recently it completed a huge water storage area on the top of the mountain into which it pumps billions of gallons of water during slack periods to use to create electric power when most needed.

The company sponsors popular bus tours of the facilities and boat trips on the adjoining Connecticut River. It employs persons who are interested in this type of environmental activity, which is carried on in cooperation with the Massachusetts Department of Natural Resources and the Division of Fisheries and Game.

Supplying electric power is probably our nation's fastest growing industry. And rural electrification which assures abundant food production is a vital part of it all.

To staff such enterprises, our hundreds of power companies need well-trained personnel. Qualified men and women are needed for business administration, for sales, for engineering— especially agricultural engineering—for community development, for public relations and customer relations, and for all the essential activities of a huge service enterprise.

Graduates in business administration, engineering, home economics, and agriculture are employed at good starting salaries and are given valuable in-service training to prepare them for advancement.

Power Company Farm Advisors

Carefully selected, well qualified, and experienced, these specialists aid farmers with services such as these:

> Furnishing competent advice on the use and installation of electric equipment.
> Helping design materials-handling systems, and feeding systems.
> Checking adequacy of power supply and wiring system.
> Advising on most economic use of electric power.
> Helping reduce labor requirements.
> Advising on selection and purchase of electric equipment.
> Cooperating with agricultural colleges and their extension service and experiment stations.
> Addressing farmers' meetings on electrical development.
> Conducting information programs by mail or by local advertising.
> Working with youth organizations.

The service of the power company's farm advisor is free. He or she is always on call to help the farmer improve operations with the efficient use of electric power.

SOME RESULTS OF RURAL ELECTRIFICATION

Rural electrification has transformed the countryside, bringing inestimable benefits to farm living and to farm production. It has made great changes in the farm home. Formerly a dimly lighted, crudely furnished, isolated dwelling, it lacked all the modern conveniences of daily living. Now the farm homemaker has as many, perhaps even more, electric "servants" as the city homemaker. She has good lighting; a pressure water system with modern plumbing; a host of power-driven kitchen and housekeeping appliances, such as a refrigerator, a food freezer, a food processor, an electric range, a dishwasher, a clothes washer, and a clothes dryer; and various communications media, including a TV, a radio, and a telephone. Many farm homemakers use electricity for more purposes than do their city counterparts.

Rural electrification has spread throughout our open country with remarkable speed. It is a development of just a few decades. But within that brief period, farm women have learned quickly and well how to use this new, beneficent power.

Today, folks "live electrically" in the farm home.

In farming operations, electric power performs countless jobs. Most important, it has lessened the drudgery of hard labor in farm chores and the hard work of handling materials. Few of us realize the tremendous weight of the materials that must be lifted, moved, and carried on the farm.

Look at the annual materials-handling job on a dairy farm for every 10 cows:

> 60 tons of silage
> 15 tons of grain
> 10 tons of hay
> 40 tons of milk
> 5 tons of bedding
> 90 tons of manure

Now, with electric power, most of this work is done by "pushing a button."

Livestock feeding has become automatic. Preparing, grinding, mixing, and measuring feed and distributing it to the animals are all done by electric power.

Our great commercial poultry industry could not function as it does today without electricity. Electronic controls make large-scale operations possible. Feeding and watering the birds, collecting, candling, and grading eggs are all done automatically. Some egg-producing concerns market over a million dozen eggs annually. And all the work is done with very few employees.

Milk production methods have changed greatly. Milking is done by machine, with the milk enclosed—not exposed to air—all the way from the cow to an electrically refrigerated bulk tank.

Thence it is transported by a refrigerated tank truck to the processing plant. At every stage, from the cow to the consumer, it is kept clean, sanitary, and fresh with electric power.

Farmstead operations are electrified and farm buildings, alive with electric power, become a real functioning unit for processing and refining farm products.

A list of all the electric jobs on the farm would take many pages. Electricity gives light for late evening chores, makes hens keep longer working hours, facilitates sanitation, kills germs, controls humidity and ventilation, keeps water pipes from freezing, controls the flow of irrigation water, warms up the tractor for easy starting on cold winter mornings. No wonder farmers use 136 billion kilowatt hours of electricity each year.

No wonder today's farms are different!

BUT FARMERS ARE OUTNUMBERED

They are a minority group in their own country. Not widely known and little published is the fact that more non-farm people than farmers live in our rural areas. In fact, about three times as many. The most recent data available show about 40 million non-farm "ruralists" compared with 13 million farm people.

Who are these non-farm ruralists? They are a cross section of American people who make their living in business, professions, trade, industry, and education just as urban people do. They prefer to live in the country along with farmers and those who have found a retirement home there. Some commute to their work in the cities which may be many miles distant. Others work in the thousands of industries that have moved their plants to rural areas since the coming of rural electrification.

Rural electric cooperatives now serve 332,000 "small" commercial loads and nearly 20,000 "large" commercial loads for rural industries; over half of their annual revenue comes from non-farm customers.

The lines that once marked limits between city and country have been erased. The walls that once separated the two have been tumbled down by rural electrification.

FIGURE 18-5. Headquarters of Consumers Power, Inc. This rural electric cooperative furnishes power for the farms, farm homes, and rural industries for territory it serves. (Courtesy, USDA)

So, if you choose a career in this industry, you will work with all kinds of "clients." Some will be farmers, some will be homemakers, and some will be business persons and industrialists. Rural electrification serves the diverse needs of many different kinds of customers.

It is a young, rapidly growing industry with lots of room for qualified, enthusiastic men and women.

HOW FAST IS IT GROWING?
WHAT IS ITS FUTURE?

Farm and rural use of electricity has doubled twice in the past 20 years. There is good reason to believe that the rapid growth will continue among farmers, but even more with non-farm rural residents and rural industries. Rural electric cooperatives added about 160,000 new customers in a recent year.

Here are some excerpts from a recent report of the Rural Electrification Administration:

> Rural distribution systems continue to require increasing supplies of wholesale power to meet the needs of existing and new customers.

Rural consumers are using ever increasing amounts of electric energy. In addition to farm uses, electric power encourages industrial and economic development of rural areas, and has created a rural market for electrical appliances and equipment estimated at $1 billion a year.

With new uses for electricity coming each year, with more complete use of air conditioning, home heating and heat pump installations, you may be sure the demand for electric power will continue to grow—and grow rapidly.

OPPORTUNITIES IN RURAL ELECTRIFICATION

Many agencies and enterprises participate actively in rural electrification. Most all of them want to employ qualified college graduates. Here's a list of some that may afford you a career opportunity:

Commercial electric utility companies
Rural electric cooperatives
U.S. Government agencies:
    Rural Electrification Administration
    Federal Power Commission

FIGURE 18-6. Rural electric cooperative offices employ many women. (Courtesy, Rural Electrification Administration)

Bureau of Reclamation
Tennessee Valley Authority and similar power au-
    thorities
Manufacturers of electric equipment and appliances
Dealers and suppliers of equipment and appliances
Food processing and other rural industry plants
Irrigation districts
Electric contractors
Associations—trade, educational, technical, promotional
Interindustry farm electric groups
Communications media:
    TV, radio, newspapers, magazines, journals
Agricultural extension service and experiment stations
Youth programs
Rural area development programs

WHOM DO EMPLOYERS WANT? WHAT TALENTS DO THEY SEEK?

A recent issue of the *University of Buffalo Alumni News* ran
the ad which follows. Could you qualify? You could if you took an
electrical engineering course in your chosen college and got some
experience by working for a utility or electrical contractor for a
year or so:

> **Staff Engineer**—Under the direction of an Area En-
> gineer, lays out and plans the construction of electric
> transmission distribution lines.
>
> **Specific Responsibilities:**
>
> – Collects and analyzes operational data such as sys-
>   tem load demands.
> – Evaluates and recommends additional facilities to
>   meet contract specifications, increased or future
>   loads.
> – Recommends shortest route to avoid interferences
>   and prepares documents for obtaining easements
>   on right of way.
> – Performs detailed calculations to draw up con-
>   struction specifications, such as cable sag, pole
>   strength, necessary grounding.
> – Directs activities and provides technical supervi-
>   sion of lower rated employees.

The ad also said you must speak, read, and write English; so
take the necessary courses.

The many agencies that participate in rural electrification render a great number of different services. To do so they need many kinds of employees, many different talents and abilities. We shall not attempt to write a definite job description of each position to be filled, but the following general classification of personnel needs may help you understand the opportunities in this agricultural industry.

Personnel Needs of the Rural Electrification Industry:

For Professional Services

Agricultural engineers
Electrical engineers
Agricultural economists
Home economists
Animal and poultry breeders
Agronomists
Agricultural research scientists
Farm managers
Managers of research farms
Farm advisors

FIGURE 18-7. Electrical engineers working on a power plant problem. (Courtesy, Rural Electrification Administration)

For Business Services

Business managers and assistants
Accountants
Auditors
Sales engineers
Equipment salespersons
Fieldpersons
Supervisors
Business machine operators
Clerks
Electricians
Maintenance workers

Other specialists for:

Communications

Journalism
Newsletters, TV, radio, bulletins, advertising
Sales promotion
Technical information

Public, Member, and Consumer Relations

Conduct and address meetings
Answer general complaints
Develop good will
Support legislative programs

Community Development

Sociologists
Home demonstration agents
Economic developers
Leaders of youth programs

Recreation Projects (A new cash crop for rural areas)

Rural area development
Construction and maintenance of recreation centers
Operation of centers

PREPARING FOR EMPLOYMENT IN RURAL ELECTRIFICATION

Our colleges of engineering, business administration, agriculture, and home economics afford the basic essential education that will qualify you to begin work in this field. Your choice of one of these professions will probably determine the area of rural electrification where you start your career. A bachelor's degree with a good college record will assure you of full consideration of "recruiters."

In-Service Training

The commercial power companies, the rural electric cooperatives, and the REA in Washington D.C., all offer in-service training. Here's a brief description of two of the programs offered by the REA, which lead to civil service positions with that government agency.

1. *Seniors in Business Administration, Accounting, or Economics:* This all-expense-paid course (for seniors or recent graduates) covers utility operation, management accounting, loan appraisal and related work. It prepares students for civil service positions, primarily as fieldpersons or accountants.

   Fieldpersons advise and assist electric and telephone cooperatives with their applications for loans from the REA and also help the locals with management and operations procedures.

   Accountants conduct preloan and postloan audits for the local cooperatives and help them establish accounting and record-keeping systems.

2. *REA Training for Electrical Engineers:* After being employed by the REA, the young engineer is given 6 to 12 months of training. Most of that time is spent in Washington, D.C., but some is spent in short field assignments with a rural electric cooperative.

   The work in Washington includes lectures and discussions supplemented by on-the-job experience. The training includes these areas of study:

   System planning, economics, and design
   Steam, diesel, hydro, atomic, gas turbine plant design
     and construction
   Transmission and distribution system design
   Substations and switching stations
   Pole line design and construction
   System protection and inductive coordination

> Radio, power line carrier, supervisory control, and tele-
> metry systems
> Power plant, transmission and distribution operation,
> and maintenance

Some of our colleges offer a separate course in rural electrification; some emphasize it in connection with other courses. You
may be able to choose electives that will give you the advantage of
such offerings.

Read again the suggestions in Chapter 4, "How to Prepare for
a Career in Agribusiness." You will advance faster if you develop
a full appreciation of the impact of rural electrification on our rural
community. Do not let yourself become isolated within your own
specialty, regardless of its importance. Courses in business administration are highly desirable even though you may be an engineer
or an agricultural scientist. Most every component of this industry
as well as other industries depends upon the volume and value of
its total sales and services.

To move upward toward the top, you must be able to take
your part in promoting the general welfare of the enterprise that
employs you.

Chapter 19

WOMEN IN AGRIBUSINESS
AND INDUSTRY

WOMEN IN AGRIBUSINESS
AND INDUSTRY

Women today are faced with many challenges, new problems, new opportunities, and new patterns of life. Few of us realize the remarkable progress they have made in business and industry. Today, about one-half of our nation's workers are women. Nearly half of them are married, many with children. Some 40 million now are at work throughout our whole economy.

In 1977, approximately 74,000 women owned or managed farms in the United States. Seventeen percent of all farm workers are women. Female enrollment in agricultural colleges has risen about 90 percent since 1973. Agriculture was traditionally a man's profession and Home Economics a woman's field; however, now women are doing the same jobs that traditionally have been done by men.

Shortage of labor and more governmental paper work in hiring labor have made women more valuable than ever in running their farms as businesses.

Many women's organizations in agriculture have been formed, and women find themselves buying equipment, owning farms, and operating them as businesses. Mechanization has made sheer physical strength less important in farm work. A woman can drive today's machinery as easily as a man can.

Almost all occupations are now open to women, due in no small measure to the strong-minded feminists who demonstrated women's right and capability to participate in the work of our nation. In a recent survey of 479 industrial occupations, our Census Bureau found women in *every one* of those occupations. So you can be sure that women can do, and do, just about everything that men do in business, industry, and agriculture.

Under existing federal law there can be no distinction made between men and women in employment. If a woman is moved into a rougher and harder job for which her physical strength is not equal to the task, neither management nor the women's local union can help her.

Women in farm organizations have been very effective in keeping up with the latest state and federal legislations. They have appeared before many congressional committees, the tax reform committee, for one example. Many more challenges await women in the field of communication concerning the problems in agribusiness.

An important milestone in women's rights was an action of our Civil Service Commission in 1962 which terminated the traditional practice of barring women from many government positions. Since then, women have used the Civil Service Entrance Exami-

FIGURE 19-1. Women find attractive careers in the cut-flower industry. Here, in San Francisco, the woman manager of a wholesale market is interviewed by a reporter from the Market News Service of the U.S. Department of Agriculture. (Courtesy, USDA Photo Service)

nation as their gateway to good positions. Now they play a larger role in government plans and programs. Today, one of every five federal employees is a woman; one of every three federal "white collar" workers is a woman. The U.S. Department of Agriculture is the largest federal employer. It affords maximum opportunity for women to advance. The federal merit system permits a woman to enjoy both a career and a home life. After three years of satisfactory service, a "career status" is established. Should a woman have to leave her job for home obligations, this status qualifies her to return to government service without another examination.

MILLIONS OF WOMEN NOW WORK
IN BUSINESS AND INDUSTRY

They hold all kinds of positions in manufacturing, in the service industries, in sales, in clerical work, in the professions, and as managers and proprietors.

Here are some of the principal areas where women work:[1]

> Management, official, and proprietor positions are held by nearly 5 percent.
> Clerical work employs over one-third of all women workers. (They have captured this area, almost entirely replacing men.)
> Sales work employs 6.8 percent.
> Retail trade employs 18 percent.
> Finance, insurance, and real estate comprise 6.2 percent.
> Public administration uses 4.6 percent.
> Transportation, commerce, and public utilities employ 3.5 percent.

Women are well represented in retail business, and nearly half of all persons in finance, insurance, and real estate are women. Each year more women obtain professional positions especially in public relations, personnel administration, communications, publicity, community development as social workers, librarians, reporters, editors, auditors, accountants, data processors, and computer programmers.

---

[1]Data from U.S. Department of Commerce.

Enrollment of Women Increases
in Graduate Schools

Recently the U.S. Department of Health, Education, and Welfare (HEW) presented the following comparisons:

	Percentage of Enrollment	
	1970–71	1977–78
Medical schools	9.6	25.6
Law schools	8.0	27.4
Other graduate schools	38.7	44.7

There is a definite need for more women in the professions. Many agribusiness companies and related industries employ several women lawyers and often a woman doctor.

Demand for Women Workers Is Increasing Rapidly

Their future is bright. In 1998 we will need over 42 million more workers and about half of them will be women. In 1976, 47 percent of all women in the United States were working. Single women made up 59 percent of the total.

Our expanding economy, new industries, new products, and activities will bring new career opportunities for women in the professions, in business administration, in manufacturing, and in communications. The need for well educated, capable workers in the highly skilled technical and professional fields is urgent. Each year more are needed; within a few years such highly trained workers will make up a large portion of all our working people.

Where can these workers be found? Among the women college graduates. The National Manpower Commission estimates that women now comprise half of the professional manpower reserve.

Indeed, their future prospects are bright. Each year, business, especially agribusiness, employs large numbers of women. They work in all the many components of agribusiness and in the industries, services, and sciences that support it. Women are "at home" in agribusiness. It is natural that they should be.

Women Started Agribusiness Centuries Ago

You can find evidence to indicate that women really "discovered" agriculture and the secret of seeds. And probably they also started the textile industry and succeeded in domesticating small animals. These were revolutionary achievements. They assured a much more dependable food supply than the occasional "kills" brought in by their hunter husbands. And when small surpluses accumulated, women food producers probably traded with their neighbors and became "distributors" as well as "producers."

But we have progressed far from those ancient days when women founded agribusiness. Today it is a highly complex enterprise employing millions of men and women possessing a great diversity of skills and talents. Women are still its enthusiastic supporters and essential participants.

There is good reason why they are. Much of agribusiness and

FIGURE 19-2. Workers in a cooperative's laboratory developing new food products. (Courtesy, Farmer Cooperative Service, USDA)

many of the industries and sciences that support it relate to the home and basic family needs. Look back over the industries we have discussed—food, dairy, meat, textile material, ornamental horticulture, rural electrification, cooperatives. In all these areas and others, women have definite interests and special talents which make the whole system function more effectively.

WHAT WOMEN DO AND WHERE MORE ARE NEEDED

Agribusiness covers such a broad field here at home and over-seas that many different kinds of workers are needed—from filing clerks to highly trained scientists and experienced executives. Today women fill all positions from the bottom to the top of the business status ladder.

Here's a general classification that shows the personnel needs of agribusiness, and the agencies and interests that serve it. Women now work in all of these areas and more are needed. And they have succeeded in their goal of "equal pay for equal work."

For Professional and Kindred Services

Economists
Home economists
Sociologists
Nutritionists
Dietitians
Chemists
Agricultural scientists:
    Horticulturists
    Botanists
    Entomologists
    Plant pathologists
Extension teachers
Educational directors
Librarians
Lawyers

For Business Administration

Executives and officers
Managers and assistants
Treasurers and controllers
Department executives
Supervisors and inspectors

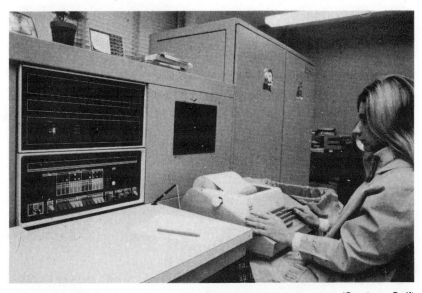

FIGURE 19-3. Many women are employed as computer programmers. (Courtesy, Swift and Company)

Auditors
Accountants and bookkeepers
Business machine operators
Computer programmers
Salespersons
Buyers
Office managers
Secretaries and typists
Cashiers and clerks
Hostesses and receptionists
Custodians

Specialists for Other Business Services

Market research
Consumer research
Business forecasting
Business analysis
Product testing
Comparative shopping
Transportation and traffic
Laboratory operations
Store decoration and product display
Personnel administration
Education

Photography
Credit and collections
Insurance

For Communications

Journalists
Technical writers
News writers and reporters
Speakers
Advertising specialists
Editors
Sales promotion specialists

For Government Service (in local, state, and federal government)

Home economists
Extension workers
News reporters
Agricultural statisticians
Food technologists
Dietitians
Laboratory technicians
Kitchen testers

FIGURE 19-4. Good "customer relations" often begin with the receptionists. Women are preferred for this important service. (Courtesy, Rural Electrification Administration)

Food and product testers
Inspectors
Market specialists
Personnel workers
Secretaries
Typists

Part-Time Jobs for Women: Many women find part-time employment highly desirable and rewarding. It often fits in well with their other obligations.

Here are some of the areas where they work:

Writing (articles and news items)
Auditing and accountancy
Temporary office work
Comparative shopping
Product testing
Consumer research

Specialists for "Good Relations": One of every four workers in the field of public relations is a woman.

Customer relations
Labor relations
Personnel administration
Employee relations
Recreation leaders
Rural area development

More Women Needed in Agribusiness: Here's a message from the California Polytechnic State University at San Luis Obispo:

Increasingly, farm related businesses are employing women in credit departments, real estate, title insurance, food processing, accounting, personnel, banking, and other related fields. Government agencies, too, seek women for these and other occupational areas including marketing, business forecasting, information, and public relations. Women who have business training are in demand for a wide variety of positions which include managerial and supervisory duties in clean and pleasant surroundings.

Cooperatives Use More Women Employees: Many cooperatives use women employees with skills as laboratory technicians, with good education in nutrition, and with knowledge of consumer demands. These workers have developed a number of new products over the years in order to give the farmers better markets for their poultry and poultry products. These new products are sold under the cooperative's brand name.

Opportunities for Home Economists: Look over some college curricula in home economics. They will tell you of career opportunities, describe the courses that can prepare you for them, and *tell what graduates do.* Such curricula require at least four years of study although some colleges offer shorter courses of one or two years. These, too, can be of great value to you and enhance your chances for remunerative employment. But the four-year courses awarding you a degree of Bachelor of Science should be your goal. Better yet, if you can, look beyond that to graduate study.

Here are a few typical excerpts from curricula in home economics at various colleges. Your own state colleges offer similar programs:

> Food and Nutrition Curriculum—prepares for food service administration in institutions such as hospitals, school lunch rooms, commercial organizations, restaurants, inns, hotels, etc.

> Graduates are working for the nation's big food equipment manufacturers—some edit magazines and home department editions of newspapers.

> Home Economics Journalism. Students write about foods. Graduates may become editorial assistants on food page editions of newspapers. Requests for home economists trained in journalism exceed the number so trained.

> Graduates may become food demonstrators for a commercial company, or on TV, or a food analyst, a food technician, a nutritionist, or conduct a test kitchen.

> ... trains home economists for extension work, home economics journalism, home equipment demonstration—selected courses prepare for industrial, hotel, and other types of food service.

> A graduate of this curriculum finds many alternatives open to her. She may become a woman's-page editor on a newspaper, or an editorial assistant on food or equipment on a magazine or in an advertising agency.

FIGURE 19-5. Here the owner poses in front of her well landscaped home. Low-growing plants were used for her one-story, rambler style home. Ornamental horticulture and landscaping offer many opportunities for qualified women specialists. (Courtesy, USDA Photo Service)

> This Food Science curriculum prepares students for leadership in the food industry—for careers in plant operation, plant management, quality control, research, and teaching.

Careers in Ornamental Horticulture and Floriculture: This area of agribusiness is particularly attractive to women. In it they may find satisfaction and real opportunity for creative achievement as well as good financial return.

Reread Chapter 13 to review the many vocational and professional opportunities in this field. Ornamental horticulture has, indeed, become a big industry. The value of our annual production of horticultural specialties is mounting toward $1 billion and a large part of this production is ornamental materials.

Our colleges offer excellent training in this field. Courses include the many phases of plant production, distribution and utilization, greenhouse and nursery management, development of landscape services, supervision of parks and recreation areas. Special emphasis is placed on the use of ornamental plants in the home and home garden.

Courses in floriculture include flower arrangement for the home, for sale, and for exhibition, and flower store management. The number of retail flower shops in the United States has increased rapidly in the past few years. Many of them are managed and several are owned by women.

WOMEN CONTRIBUTE MUCH TO AGRIBUSINESS

Each year more women are employed in agribusiness. They serve in all of its many components—in business administration, in the professions and sciences that support it, in the governments that serve it, in communications that tell its story, and above all, in the channels through which its merchandise flows.

Women in agribusiness promote something they believe in. They work for their families, their homes, and their country. And

FIGURE 19-6. Many women work in research. Here, corn is being cross-pollinated for development of new varieties. (Courtesy, Agway Inc.)

now that one-half of our nation's workers are women drawing a pay check, we can see a bright future for women in business and industry—especially in agribusiness. In agribusiness they return to the basic food-producing industry founded long ago when women first "discovered" agriculture.

Chapter 20

AGRIBUSINESS AND YOU

AGRIBUSINESS AND YOU

SHOULD YOU CONSIDER A CAREER IN AGRIBUSINESS?

Ask yourself that question now that you have some information about opportunities in agribusiness. If you feel a real interest in the broad field, you will want to give it still more thought and study.

Here are some suggestions and questions that may help you, no matter what career you may choose.

Would You Like the Work?

This is an easy question to ask and a very important one. Liking your work is almost essential to success:

> It smooths your path—makes hard work easier.
> It helps you develop and grow.
> It expands your interests.
> It gives you a lively curiosity.
> It helps you recognize opportunities ahead.

Would the Work Bring You Satisfaction?

Would agribusiness give you the challenges, opportunities, and satisfaction in your working life that mean so much to your happiness?

Arthur J. Goldberg, when Secretary of Labor, gave this advice on choosing a career:

> Do not let salary consideration or the availability of certain kinds of work be your sole guides in choosing a

job. It is more important to find the kind of work that is suited to your aptitude, training, and interests. Obviously, you will not succeed in a job that does not interest you, even if it pays a higher starting salary than others and is easier to obtain.

Is Agribusiness Expanding or Contracting? What Are Its Future Prospects?

You can find many answers, and almost all of them will show that agribusiness is expanding. In the future it must furnish greater services here at home and throughout the world.

Agribusiness will grow because every day it will have more customers. In the United States today we have about 220 million people. We shall probably have 235 million by 1985. In the world today we have 4.1 billion people, and we shall probably have 4.5 billion by 1985.

No other business has, or ever can have, as many customers as agribusiness. Our grain is one example of "merchandise of agribusiness" that flows to hundreds of millions of customers around the world. Today, our country and Canada export about 40 million tons of grain a year. Experts predict much greater foreign demand in the years ahead—perhaps 100 million tons in the year 2000.

Even that huge amount will not be nearly enough to meet the world needs, because population increases so fast. But just this one example helps us see how rapidly this business must grow in your lifetime. We shall soon need more farm production, more food and fiber processing, and many more trained business specialists to control and direct the flow of food and fiber products to customers everywhere.

Few, if any, business enterprises have such bright prospects for future growth.

Examine the following listing to locate the jobs with the best future for you.

Opportunities for Employment Through 1985
for Some Agribusiness Occupations[1]

	Estimated Employment 1976	Average Annual Openings to 1985	Growth Compared to Average
Accountants	865,000	51,500	faster
Bank officers, managers	319,000	28,000	faster
Buyers	109,000	5,700	slower
Credit managers	53,000	2,500	slower
Purchasing agents	192,000	13,800	faster
Programmers	230,000	9,700	faster
Systems analysts	160,000	7,600	faster
Computer operators	565,000	8,500	average
Agricultural engineers	12,000	600	faster
Chemical engineers	50,000	2,100	average
Electrical engineers	300,000	12,800	average
Mechanical engineers	200,000	9,300	average
Actuaries	9,000	500	faster
Statisticians	24,000	1,500	faster
Biochemists	12,700	600	average
Chemists	150,000	6,300	average
Dietitians	45,000	2,800	average
Economists*	115,000	6,400	faster
Engineering & science technicians	586,000	29,000	faster
Forestry technicians	11,000	600	faster
Soil conservationists	7,500	400	average
Foresters	25,000	1,100	average
Home economists*	141,000	6,100	average
Newspaper reporters	40,500	2,100	slower
Radio, TV announcers	26,000	1,300	slower
Personnel directors, labor relations managers	335,000	23,000	faster
Insurance agents, brokers	490,000	27,500	average
Wholesale-trade salespersons	808,000	41,000	average
Manufacturers' salespersons	362,000	17,600	average
Government inspectors	115,000	7,900	slower
Social workers	330,000	25,000	faster

*Need advanced degree.

[1]*Occupational Outlook Handbook* (1978–79), U.S. Bureau of Labor Statistics.

Can You Prepare and Qualify for Agribusiness?

Check Your Personal Aptitudes and Interests: First, try to determine whether you would like the work. Talk with people in an agribusiness industry. There are some near you. They will help when they know why you are interested. They may have trade journals that describe their particular industry.

Discuss Your Career Plans with Friends and Counselors: Talk about your career plans with parents, teachers, and friends and make helpful contacts. Maybe you will also learn to be a good listener—another very valuable ability.

COLLEGE EDUCATION IN AGRIBUSINESS

Visit a College That Offers Agribusiness Curricula

Talk with the faculty members. Find out what their graduates are doing and how they like their work. Get catalogs describing the agribusiness curricula, and read about the many specialized careers for which you might prepare yourself.

Check yourself on these personal qualities and abilities. They are important in business. If you are not satisfied perhaps you will make a special effort to improve.

How do you rate yourself on the following?

> *Reading* – speed and comprehension
> *Communications* – ability to express your ideas clearly and forcefully, orally and in writing
> *Public relations* – ability to make friends and to participate in school activities
> *Cooperation* – ability to work well with others, ability to follow as well as to lead
> *Courage* – willingness and confidence in accepting responsibilities and assignments, readiness to tackle problems and to try to solve them

FIGURE 20-1. Beginning college student and her advisor discuss her course of study. (Courtesy, Cornell University)

Four Years of Serious College Study Are Desirable

Can you qualify for college entrance? Are your high school grades good enough? Can you find ways to pay for college education? On these problems, as on so many others, your counselor will help greatly. He or she will help check college entrance requirements, admission tests, probable costs, available scholarships, work opportunities, and other essential information.

Be prepared to study hard; there are no "snap" courses in agribusiness.

You will find more data on the cost and the value of college education later in this chapter. A degree is highly desirable for a professional career in agribusiness.

Every year, our expanding economy places a greater premium on management talent. Demand increases for trained young persons to fill management positions. But a first requirement is college education. Without a college degree, you would have less chance to reach management or administrative status.

Agribusiness Jobs Requiring a Four-Year College Degree

Systems analysts	Soil conservationists
Bank officers	Engineers
Actuaries	Geophysicists
Underwriters	Meteorologists
Accountants	Biochemists
Advertising workers	Soil scientists
College student counselors	Mathematicians
Hotel managers	Statisticians
Industrial traffic managers	Chemists
Market research workers	Food scientists
Personnel workers	Dietitians
Public relations workers	Economists
Purchasing agents	School counselors
Health inspectors	Employment counselors
Teachers	Cooperative extension
Manufacturers'	workers
salespersons	Home economists
Securities salespersons	Industrial designers
Airline dispatchers	Landscape architects
Foresters	Newspaper reporters
Range managers	Technical writers

What Does College Education Cost?

Total costs depend pretty much on you—your habits and desires and how you live while in college. And costs vary substantially in different regions of our country. Costs at private colleges are higher than at publicly supported colleges.

As a guide in planning, figure $2,500 as your annual cost at a publicly supported college in your own state. That amount may cover tuition, required fees, room and board, books and supplies and miscellaneous expenses. Expenses for out-of-state students are more because tuition is higher.

You can get complete information on costs by writing to the dean of agriculture at your state college.

Possibly you can attend a junior college or a branch of your state university near your home for a year or more. That might reduce your total costs. Discuss this with your counselor. Perhaps you can take subjects and earn credits there that would fulfill

some of the requirements for your baccalaureate degree in agribusiness.

Some Typical Recent Tuition Charges for One Year

East	Dollars
Boston University	4,230
University of Connecticut, Storrs	1,600*
Georgetown University, Washington, D.C.	4,100
Princeton University	5,100
State University of New York	750*

South	
University of Arkansas, Fayetteville	460*
Baylor University, Waco, Texas	1,650
Duke University, Durham, North Carolina	3,830
Florida State University, Tallahassee	700*
Morehouse College, Atlanta	1,950

Midwest	
Antioch College, Yellow Springs, Ohio	4,088
University of Cincinnati	710
Indiana University, Bloomington	870*
University of Missouri, Columbia	678*
Washington University, St. Louis	4,300

West	
California Institute of Technology	4,338
University of California, Los Angeles	702*
University of New Mexico, Albuquerque	576*
Reed College, Portland, Oregon	4,430
University of Washington, Seattle	687*

*Public universities charge more for out-of-state stu-
dents.

Note: Figures given do not include food or housing or
fees which could be $1,500 to $4,000 more.

Can You Get Financial Assistance?

Yes. Financial assistance to students is increasing every year.
Loans, scholarships, and part-time jobs are available. You may be
able to qualify for some such aid that will pay a large part of your
costs. Student loans come from many sources.

Federal Government: The U.S. Office of Education supports

three types of financial aid for college students: *grants* which are not repaid, *loans* which are repaid after graduation, and a *work and study* program.

There are five programs of financial aid:

1. *Basic Educational Opportunity Grant (BEOG) Program:* You can obtain an application from your counselor or write to BEOG at either P.O. Box 84, Washington D.C. 20044, or P.O. Box B, Iowa City, Iowa 52230.

2. *Supplemental Educational Opportunity Grant Program:* You apply through the college financial aid officer of the college you attend.

3. *National Direct Student-Loan Program:* You apply for a loan through the financial aid officer of the college you attend.

4. *Guaranteed Student-Loan Program:* You borrow directly from your bank, credit union, or saving and loan association. Ask for information at your bank.

5. *College Work-Study Program:* This program provides jobs while you are in college. Apply through the financial aid officer of the college you attend.

You may obtain more details on these programs by writing to: Office of Education, HEW, Washington, D.C. 20202, and asking for Publication No. (OE) 77-17907 or its replacement. Check with your counselor first.

Banks and Financial Institutions: Each year banks, finance companies, and life insurance companies are increasing loans to students. Interest rates run somewhat higher than on government loans, but the amount of the loan may be greater—enough for four years of college, with funds supplied as needed. Many banks have special provisions to enable parents to borrow for their children's education.

College Loan Funds: Most colleges have funds of their own which are loaned to qualified students at relatively low interest rates.

In making a loan to you—whether from its own funds, federal funds, or other monies under its jurisdiction—the college will consider your ability, your record, and your need. You will be asked how much your family can pay, what savings are available, and how much you may be able to earn during your college years.

You are not likely to get a loan large enough to pay your total costs for four years. Loans can help, but they must be supplemented from other sources.

Scholarships Come from Many Donors: Each year some 500,000 scholarships are available to college students. They average about $450 per year in value, for a total of close to $225 million.

Scholarships come from our federal government, our states, business, industry, banks, foundations, associations, individuals, and colleges and universities. Many agribusiness firms offer scholarships.

How Can You Find Out About Scholarships? Your high school guidance counselor and principal stand ready to help you in this problem as in other problems. They will have much pertinent information on hand. Take full advantage of their helpful counsel.

College catalogs describe the scholarships offered at their respective institutions. Some colleges may send you a separate listing of available scholarships. This may give the names of donors, specify the value per year, tell how the recipient is selected, list requirements for eligibility, and explain how the funds are paid to the recipient. You might receive a questionnaire on which you could present your qualifications.

Visit a college and ask at the information office about available scholarships.

The American Legion will send you a most helpful and valuable booklet entitled "Need a Lift?" which gives a very comprehensive list of scholarships and contains much other good information and counsel for you.

Can You Qualify for a Scholarship? First you will need to learn the conditions which govern awards. These vary greatly. In some cases the conditions are established by the donors. Some scholarships are restricted and designed for a definite purpose or goal. Others are more general.

Some basic rules and requirements may be set up, but often exceptions are made. Usually awards are made on the basis of ability and need. You would probably have to show an excellent record scholastically and be able to prove that financial assistance is essential.

Scholarships and *awards* are available at most colleges and universities, and students generally apply for them. Two examples are:

> The College of Agriculture, University of Arizona, has 55 scholarships and 10 awards available to its students. The amounts vary from $50 to $1,500, averaging about $400.
>
> The College of Agricultural and Life Sciences, The University of Wisconsin, has about 135 awards and scholarships for its students. Student loans are also available.

Can You Work Part-Time While in College?

Again, the answer is yes. Every college has set up an office to help students find part-time jobs. Local business concerns cooperate with the college in providing employment opportunities for students. Various government agencies also offer part-time jobs.

Summer employment has substantially helped many students to finance their college study.

College catalogs describe job opportunities available and will aid you in obtaining such employment.

Some technical schools, such as the Rochester Institute of Technology at Rochester, New York, alternate work with study.

How Much Is a College Education Worth?
What Is Its Value?

You would find it difficult to measure its value in dollars. Much of its real value cannot be expressed in financial terms. How it can broaden and enrich your life, expand and stimulate your interests, sharpen your abilities, give you competence and confidence, and bring satisfaction and happiness—all these values can't be measured in dollars. Actually its full value depends on you and

FIGURE 20-2. Will you be in a graduating class too in a few years? (Courtesy, S. Pendrak, SUNY–Cobleskill)

how you use and apply your advanced education to all facets of your life.

A person investing money in stocks and bonds rather than investing in college would only realize 5 to 8 percent return, while money used for college could bring a return of 9 to 14 percent in increased income.

Although college remains a financially rewarding investment for most graduates, the fringe benefits of higher education—better jobs, happier homes, and generally fuller lives—are even more significant.

Dr. Seymour Harris, of Harvard University, has estimated that a college graduate should have a $300,000 advantage over a non-college graduate during the working years. And the college graduate has an advantage all through his or her working life. During middle age the income of the non-college person tends to decrease, while that of the college person continues to increase.

What If You Don't Have a College Education?

Today, without college training, you will be at a disadvantage. Most of our agribusiness firms hire only college-trained persons (except for factory workers). Company representatives seeking new staff employees confine their interviews to students who have completed, or soon will complete, their college education.

Without college training you will get much less consideration from prospective employers. The best starting positions are usually offered to graduates. A recent year's graduates of one midwestern college of agriculture got average starting salaries of $10,900 per year. Nearly all were employed in non-farm areas of agriculture and in agribusiness.

Of course, commencement day does not mark the end point of your business education. After graduation comes in-service training within a specific agribusiness industry. There you will follow programs designed to give you experience and an opportunity to learn, to grow, and to progress to managerial status.

College Education Pays!

It pays to the end of your working life. And it pays when you begin your career.

College education opens the gate to employment. It means that prospective employers will be interested in you. They know your formal education has given you basic fundamentals of economics and business practices. They know you have "tools" to work with; they know you possess abilities and interests that are essential in professional business management.

Chancellor Clifton R. Wharton, Jr., of the State University of New York, had this to say concerning the worth of a college degree:

> U.S. Department of Labor statistics show that, during one recent month, 7.2 percent of the U.S. work force was unemployed. Among those persons with a high school diploma, 8.2 percent were unemployed, while 6.3 percent of those with one to three years of college were unemployed. What about college graduates? Only 2.8 percent of the work force with four or more years of col-

lege were unemployed. Thus, those who finish college with a degree are 50 percent more likely to be employed than those who drop out before graduation. And they are three times more likely to be employed than high school graduates.

College education pays. It pays in dollars, and it pays in the more important ways that will make your life richer and fuller. You will recover your costs many times over.

No matter what career you choose or what your life's work may be, a college education is perhaps the best investment you can ever make.

INDEX

INDEX